AF601236

Question Bank on Agricultural Extension

The Authors

Dr. B.B. Singh (M.Sc., Ph.D., PGDEM, FISEP, FLWS) obtained M.Sc. in 1992 from Dr. R.M.L. Avadh University, Faizabad and Ph.D. in 1996 from D.D.U. Gorakhpur University, Gorakhpur. Dr. Singh has worked in the capacity of S.R.F. (ICAR), Research Associate (UGC) and Faculty at N.D. University of Agriculture and Technology, Kumarganj, Faizabad (U.P.). He was actively engaged in teaching and research at B.Sc. (Ag.), M.Sc. (Ag.) and Ph.D. levels. Dr. Singh is the first recipient of 'Young Scientist Award' by International Society for Environmental Protection, India in the year 1997. He has published several research papers in journals of repute and also authored more than 24 books. Presently, he is actively involved in writing and editing of books.

Dr. C. Karthikeyan completed his M.Sc.(Ag.) and Ph.D., in Agricultural Extension during 1992-97 at Tamil Nadu Agricultural University (TNAU), Coimbatore. Graduated MBA (HRM) during 2010 at Bharathiar University, Coimbatore. He started his career as Scientist (Agril. Extension) at CIFT, ICAR, Cochin and later joined as Assistant Professor (Agril. Extension) at TNAU, Coimbatore. Currently, working as Professor (Agril. Extension) and Coordinator of KVKs in TNAU, Coimbatore. He published 16 standard books, 14 book chapters, 16 technical books, 104 scientific and popular articles in International and National journals of the repute.

Dr. Farooq Ahmad Aga acquired Doctorate degree in Agronomy from Sher-e- Kashmir University of Agricultural Sciences and Technology, Kashmir (SKUAST-K). Dr. Aga has an experience of 30 years in Research and Extension. Besides teaching, Dr. Aga has worked in different capacities like Agricutural Extension Officer, Sub-Divisional Subject Matter Specialist in Department of Agriculture, Govt. of Jammu and Kashmir and as Asst. Professor, Chief Warden, Training Associate in KVK. Currently, he is working as an Associate Professor and Programme Cordinator of KVK in Directorate of Extension SKUAST-K. Dr. Aga has rendered a commendable service in Agricultural Extension and is a prolific writer broadcaster and Telecaster in Krishi Darshan and Radio. Dr Aga is Editor of Urdu Extension Journal Zarie Paigam besides Editor of Research Journal of Agricultural Sciences in English.

Question Bank on Agricultural Extension

B.B. Singh
C. Karthikeyan
F.A. Aga

Daya Publishing House®
A Division of
Astral International (P) Ltd.
New Delhi 110 002

Publisher's Note:

Every possible effort has been made to ensure that the information contained in this book is accurate at the time of going to press, and the publisher and author cannot accept responsibility for any errors or omissions, however caused. No responsibility for loss or damage occasioned to any person acting, or refraining from action, as a result of the material in this publication can be accepted by the editor, the publisher or the author. The Publisher is not associated with any product or vendor mentioned in the book. The contents of this work are intended to further general scientific research, understanding and discussion only. Readers should consult with a specialist where appropriate.

Every effort has been made to trace the owners of copyright material used in this book, if any. The author and the publisher will be grateful for any omission brought to their notice for acknowledgement in the future editions of the book.

Cataloging in Publication Data--DK
Courtesy: D.K. Agencies (P) Ltd. <docinfo@dkagencies.com>

Singh, B. B. (Agriculturist), author.
Question bank on agricultural extension / B.B. Singh, C. Karthikeyan, F.A. Aga.
pages cm
ISBN 9789390384440 (Int. Edition)

1. Agricultural extension work--India--Examinations, questions, etc. I. Karthikeyan, C. (Chandrasekaran), 1970- author. II. Aga, F. A. (Farooq Ahmad), author. III. Title.

S544.5.I4S56 2017 DDC 630.715 23

Published by : **Daya Publishing House®**
A Division of
Astral International Pvt. Ltd.
– ISO 9001:2015 Certified Company –
4736/23, Ansari Road, Darya Ganj
New Delhi-110 002
Ph. 011-43549197, 23278134
E-mail: info@astralint.com
Website: www.astralint.com

भारतीय कृषि अनुसंधान परिषद
कृषि अनुसंधान भवन-II, पूसा, नई दिल्ली 110 012
INDIAN COUNCIL OF AGRICULTURAL RESEARCH
KRISHI ANUSANDHAN BHAVAN-II, PUSA, NEW DELHI 110 012

डा. नरेन्द्र सिंह राठौड़
उप महानिदेशक (कृषि शिक्षा)
Dr. Narendra Singh Rathore
Deputy Director General (Agril. Edn.)

Phone : 011-25841760 (O)
Fax : 011-25843932
E-mail : ddgedn@gmail.com
nsrdsr@gmail.com
Website: www.icar.org.in

Foreword

Agricultural extension is the application of scientific research and new knowledge to agricultural practices through farmer education. The extension encompasses a wider range of communication and learning activities organized for rural people by educators from different disciplines. including agriculture, agricultural marketing, health, and business studies. Agricultural extension is considered to be a special branch of rural extension dealing with several economic and social aspects of farming community. In India, most of the farmers are small and marginal, they lack direct access to developing agricultural technology. Extension education aims at changing the outlook and attitude of the farming community by educating them about different modern farming practices and technologies for enhancing productivity and hence more production.

The extension process is concerned with communicating the technology of scientific agriculture to the farmers in order to transform traditional level of agriculture to better for improving their economic conditions. Hence, extension in agriculture would mean stretching out the knowledge to farmers on adoption of the new technology and improved practices in various sub-sectors like crop production, livestock rearing fodder production, sericulture, bee-keeping, horticulture *etc.*

The book entitled "Question Bank on Agricultural Extension" is a comprehensive compilation, in which questions are substantially acquired from different competitive exams. A wide coverage of entire subject divided in nine chapters *viz.* Introduction to Agricultural Extension, Agricultural Communication, Media and Information Technology, Programme, Planning and Rural Development in Agriculture, Management in Agricultural Extension, Training and Research Methodology, Adult Education, Rural Sociology and Educational Psychology, Entrepreneurial Development and Elementary Statistics. This book consists of chapter-wise objective type questions like Multiple Choice, Fill in the Blanks and True or False will enable the readers to update their knowledge and functional skill quickly. Glossary is given at the end for ready reference and information.

I believe, this book will be very useful for the students appearing for various competitive examination, like post-graduate SAUs Entrance, Ph.D. JRF, ARS, ICAR- NET, SET, Agricultural Officer, Bank (PO) (Agriculture), ADO and Allied Agricultural Examinations conducted by different agencies.

The book "Question Bank on Agricultural Extension" is a noble initiative taken by authors. I congratulate authors for preparing this comprehensive and informative compilation, which will be definitely useful to students appearing in various competitive examination.

(Dr. N.S. Rathore)

Preface

The agriculture still continue to be known as the engine of the national prosperity and development. The rapid and balanced growth of agriculture is thus essential not only to achieve self-reliance at national level but also for household food security and to bring about equity in distribution of income and wealth resulting in rapid reduction in poverty among the rural poor.

Extension education has now developed as a full-fledged discipline, having its own philosophy, objectives, principles, methods and techniques which must be understood by every extension worker and others connected with the rural development. It might be mentioned here that extension education, its principles, methods and techniques are applicable not only to agriculture but also to veterinary and animal husbandry, dairying, home science, health, family planning *etc.*

The agricultural extension system, which was primarily, meant to transfer the new techniques and new knowledge from the research sector to the farmers. In India farming community residing in rural villages need to be supported with information, knowledge and the key skill to adopt improved technologies that would enhance productivity, employment opportunities and sustainability.

The book entitled "Question Bank on Agricultural Extension" is a compilation work and questions are substantially acquired from different competitive exams. A wide coverage of entire subject divided in nine chapters *viz.* Introduction to Agricultural Extension, Agricultural Communication, Media and Information Technology, Programme, Planning and Rural Development in Agriculture, Management in Agricultural Extension, Training and Research Methodology, Adult Education, Rural Sociology and Educational Psychology, Entrepreneurial Development and Elementary Statistics.

This book consists of chapter-wise objective type questions like Multiple Choice, Fill in the Blanks and State True or False will enable the readers to update their knowledge quickly. Glossary are given at the end for ready reference.

This book would be very useful for those appearing for various competitive examination, like post-graduate, SAUs Entrance, Ph.D. JRF, ARS, ICAR- NET, SET, Agricultural Officer, Bank (P.O) (Agriculture), ADO and Allied Agricultural Examinations conducted by different agencies.

We believe that this book will be also useful for fellow teachers, researchers, extension workers and development officers for references and easy answering of many complicated questions. Every possible effort has been made to accommodate previous year's questions from various competitive examination on memory basis.

We are grateful to all those persons as well as various books, manuals, periodicals, magazines, newsletters, journals *etc.* that helped in the preparation of this book.

In spite of the best efforts, it is possible that some errors may have occurred into the compilation and editing of the book. Further queries, suggestions and criticisms for improvement of the book are always welcome and shall be thankfully acknowledged. Last but not the least, it is a pleasure for us to extend our sincere thanks to Mr. Anil Mittal, Astral International Pvt. Ltd. and his team members for his keen interest shown in preparation and publishing of this book so efficiently and promptly.

Authors

Contents

CHAPTER 1

INTRODUCTION TO AGRICULTURAL EXTENSION

Multiple Choice Questions

1. **The origin of the word extension is tranced from**
   (*a*) Greek (*b*) Latin
   (*c*) Hindi (*d*) Persian

2. **Who is the father of extension education ?**
   (*a*) F.L. Bryne (*b*) Paul Legans
   (*c*) James Stewatt (*d*) Edgar Dale

3. **Who has the first used the word extension?**
   (*a*) Vorhees (*b*) Edgar Dale
   (*c*) Seaman A Knapp (*d*) Paul Legauns

4. **Who is the father of extension education in India?**
   (*a*) K.N. Singh (*b*) J.B. Chitambar
   (*c*) O.P. Dhama (*d*) Adivi Reddy

5. **The term 'extension education' was first coined in**
   (*a*) UK (*b*) India
   (*c*) Netherlands (*d*) USA

6. **Principle of extension education is**
   (*a*) Learning by doing (*b*) Learning by seeing
   (*c*) Learning by reading (*d*) None of these

7. **Goal of extension education is**

   (*a*) Promote income of farmers (*b*) Promote production of crop

   (*c*) Promote new crop (*d*) None of these

8. **Correct sequence of steps in extension teaching is**

   (*a*) Attention - Interest - Desire - Conviction - Action - Satisfaction

   (*b*) Action - Attention - Interest - Conviction - Desire - Satisfaction

   (*c*) Desire - interest - Attention - Conviction - Action - Satisfaction

   (*d*) None of these

9. **The correct sequence sequence of innovation -decision process or adoption process is**

   (*a*) Interest - awareness- evaluation - trial - adoption

   (*b*) Evaluation - awareness - Interest - trial - adoption

   (*c*) Awareness - Interest - evaluation - adoption -trial

   (*d*) Awareness - Interest - evaluation - trial - adoption

10. **Normal rate of adoption requires how many years from the introduction of the innovation to its adoption thought the community**

    (*a*) 3-6 (*b*) 6-10

    (*c*) 9-13 (*d*) 15-18

11. **The concept of bio-village was conceived by**

    (*a*) M.S. Swaminathan (*b*) V.L. Chopra

    (*c*) K.L. Chadha (*d*) G. Perumal

12. **Extension worker is known as Rural Agent is**

    (*a*) Peru (*b*) Netherlands

    (*c*) USA (*d*) Mexico

13. **The book Rural Development: Principles and Management is written by**

    (*a*) Katar Singh (*b*) R.P. Singh

    (*c*) Daulat Singh (*d*) K.L. Chadha

14. **In which particular technology the technology of self out smarted extension to reach farmers**

    (*a*) Agroforestry (*b*) Watershed Technology

    (*b*) Hybrid Rice Technology (*d*) System of Rice Intensification

15. Social Norm is

(*a*) The moulding process of individuals and their behaviours

(*b*) The rule that a group uses for appropriate values, beliefs, attitudes and behaviours.

(*c*) The established procedures in a civilized society.

(*d*) The unwritten rules that governs the members of society.

16. The word extension is derived from

(*a*) Greek (*b*) Latin

(*c*) French (*d*) English

17. The book authored by O.P. Dhama and O.P. Bhatnagar

(*a*) Extension education and communication for development

(*b*) Education and agricultural communication for development

(*c*) Education and communication for development

(*d*) Agricultural extension and communication for development

18. Who is called the father of "University extension"?

(*a*) G. Thorne (*b*) J. Stewatt

(*c*) Seaman A. Knapp (*d*) L.D. Kelsey

19. Innovation decision period is the ______ period during which the new idea ferment in the minds of the people

(*a*) Production (*b*) Gestation

(*c*) Mental (*d*) None of the above

20. The basic unit for extension service in USA is

(*a*) State (*b*) County

(*c*) Country (*d*) District

21. August Comte is considered as father of

(*a*) Anthropology (*b*) Sociology

(*c*) Rural Extension (*d*) Village Studies

22. S.K. Dey was associated with

(*a*) Grow more food compaign

(*b*) National Adult Education Programme

(*c*) Community Development Programme

(*d*) Planning Commission

23. The cooperative extension service is a branch of

(*a*) Land Grant College System

(*b*) State Department of Rural Development

(*c*) State Agriculture Department only

(*d*) None of the above

24. State agriculture defense committee was organized in New York state of USA during

(*a*) US-French War

(*b*) US-German War

(*c*) 1st World War

(*d*) 2nd World War

25. The 4-H clubs are functioning in USA, 4-H implies

(*a*) Head, Hands, Heart, Health

(*b*) Head, Home, Health, Hands

(*c*) Home, Heart, Hearth, Head

(*d*) Hands, Health, Hope, Heart

26. The total number of adopter categories is

(*a*) 8

(*b*) 5

(*c*) 4

(*d*) 6

27. The difference between a demonstration an experiment is that

(*a*) A demonstration purposes to show a known truth whereas an experiment is a search for truth

(*b*) An experiment purposes to show a known truth whereas a demonstration is a search for the truth

(*c*) Both demonstration and experiment are search for truth

(*d*) None of these

28. The type of demonstration used to show how to carry out a practice is

(*a*) F.L.D.

(*b*) Method demonstration

(*c*) Focused demonstration

(*d*) None of these

29. A method of teaching designed to show by example the practical application of an established fact or a group of related fact is

(*a*) Result demonstration

(*b*) Method demonstration

(*c*) Feasibility demonstration

(*d*) Front line demonstration

30. County is the unit of extension work in

(*a*) Mexico

(*b*) India

(*c*) Philippines

(*d*) USA

31. Community Development Programme failed in India due to

(a) Lack of finance (b) No clear cut objectives

(c) Lack of peoples participation (d) All of these

32. The book 'Extension Traditional Media in India' is written by

(a) M.S. Swaminathan (b) S.K. Jha

(c) S.D. Parmar (c) D. Jha

33. Hosagabad model of extension emphasis on

(a) Public-Private Partnership (b) Extension by NGOs

(c) Extension Services by KVKs (d) Extension by Women SHGc

34. TRYSEM was launched in

(a) 1982 (b) 1988

(c) 1979 (d) 1981

35. IRDP was launched in

(a) 1979 (b) 1978

(c) 1982 (d) 1980

36. Diffusion is subset of

(a) Communication (b) Teaching

(c) Learning (d) Motivation

37. The gap between what is what ought to be is known as

(a) Necessity (b) Demand

(c) Comfort (d) Need

38. The author of reference book "Cooperative Extension Work" is/are

(a) Leagans (b) Kelsey and Hearne

(c) Reddy (d) F.L. Bryne

39. The author of the book " Diffusion of innovations" is

(a) O.P. Dhama (b) O.S. Rathore

(c) H.C. Sanders (d) E.M. Rogers

40. How many innovators are found in adoption categories?

(a) 2.5 per cent (b) 18 per cent

(c) 19 per cent (d) 16 per cent

41. Lab to land was part of

(a) ICAR Golden Jubilee Year

(b) The Celebration of Jai Jawan and Jai Kisan Movement

(c) Birth anniversry of Charan Singh

(d) Ministry of Agriculture Silver Jubilee year

42. Anand model for rural development focuses on

(a) Honey
(b) Soil and water conservation
(c) Milk
(d) Self-help groups

43. Journal of Rural Development focuses on

(a) New Delhi
(b) Gandhigram Rural University
(c) Nagpur
(d) Hyderabad

44. The effective approach of extension education is

(a) Method demonstration
(b) Farm and home visit
(c) Result demonstration
(d) Lecture

45. An idea, practice or object perceived as new by an individual is called

(a) Invention
(b) Innovation
(c) Intervention
(d) None of these

46. Theory of human behaviour in diffusion process was proposed by

(a) Wilkening
(b) Singh
(c) Rogers
(d) Haver

47. The disequilibrium theory was given by

(a) Rogers
(b) Lewin
(c) Johanson
(d) None of these

48. The term" innovation-decision process" was given by

(a) Rogers
(b) Rogers and Shoemaker
(c) S.W. Swanson
(d) Adivi Reddy

49. Reinforcement of "Innovation decision," is made at

(a) Implementation stage
(b) Confirmation stage
(c) Decision stage
(d) None of these

50. "Junior Farmer Clubs" are the youth clubs functional in

(a) USA
(b) China
(c) Australia
(d) Germany

51. Extension worker is called "Agricultural Representative in

(*a*) China (*b*) India

(*c*) Japan (*d*) Canada

52. Extension work in Japan is carried out at

(*a*) Village level (*b*) National level

(*c*) Prefectural level (*d*) All of these

53. Extension worker is called "Rural Agent" in

(*a*) Peru (*b*) Mexico

(*c*) Russia (*d*) India

54 MANAGE Extension Review is from

(*a*) Hyderabad (*b*) Hissar

(*c*) New Delhi (*d*) Almora

55. A History Farming System Research is written by

(*a*) Michael P. Collinson (*b*) Ogburn

(*c*) Theodure (*d*) None of these

56. The book written by Lyntom and Uday Pareek is

(*a*) Human Resource Development

(*b*) A Handbook of Training

(*c*) Training Modules

(*d*) Training for Development

57. Private Extension is mostly

(*a*) Voluntary Service (*b*) Free Extension

(*c*) Paid Extension (*d*) All the above

58. Kissan Call Centre (KCC) was launched in

(*a*) 2005 (*b*) 2004

(*c*) 2001 (*d*) 2000

59. The process through which we arrive at expectations and anticipations of internal psychological state of man is

(*a*) Psychological observation (*b*) Empathy

(*c*) Exploration (*d*) Empowerment

60. An innovation has relatively high degree of complexity, its adoption will be

(*a*) Slow (*b*) Quick

(*c*) Fast (*d*) Can't be said anything

61. Innovation- decision process is a/an

(*a*) Confirmatory process (*b*) Physical process

(*c*) Mental process (*d*) Action process

62. The reinforcement of innovation decision made occurs at

(*a*) Persuasion function (*b*) Knowledge function

(*c*) Decision function (*d*) Confirmation function

63. The first line extension system mainly comprises of

(*a*) ICAR and SAUs

(*b*) NGO

(*c*) Ministry of Agriculture and Cooperation

(*d*) Ministry of Rural Development

64. The ICAR established a section of Extension Education at its headquarters in

(*a*) 1961 (*b*) 1967

(*c*) 1971 (*d*) 1977

65. The cost of production in National Demonstration was

(*a*) Lower (*b*) Higher

(*c*) Same (*d*) None of the above

66. Training and Visit system was implemented in

(*a*) 1980 (*b*) 1984

(*c*) 1974 (*d*) 1882

67. Father of communication an is

(*a*) Aristole (*b*) Berlo

(*c*) Schhramm (*d*) W.N. Newman

68. The chairman of NITI aayog is

(*a*) Prime Minister (*b*) President

(*c*) Finance Minister (*d*) Governors of RBI

69. Who said "India lives in its Village" ?

(a) Dr. M.S. Swaminathan

(b) Jawaharlal Nehru

(c) Chaudhury Charan Singh

(d) Mahatma Gandhi

70. In 1957 Ensminger said extension is

(a) Service
(b) Training
(c) Education
(d) Transfer of Technology

71. The purpose of introducing the agricultural universities in India is to improve the standard of agriculture through

(a) Extension and Research

(b) Extension, Teaching and Research

(c) Extension

(d) Teaching

72. National Demonstration were the first line demonstration which were conducted on the farmers' field by

(a) BDO
(b) Gram Pradhan
(c) Extension Worke
(d) Research Scientists

73. The farmer who adopts the new technology just after learning about it is called a/an

(a) Late adopter
(b) Early adopter
(c) Laggard
(d) Innovator

74. The basic village institutions are

(a) Panchayat, hospital and school

(b) Panchayat, cooperatives and school

(c) Youth club, cooperatives and school

(d) None of these

75. Steps of extension teaching are

(a) Source, message, channel, recommendation

(b) Attention, interest, desire conviction, action and satisfaction

(c) Awareness, interest, trial and evaluation

(d) Tribal and evaluation

76. The result demonstrations were initiated by Dr.Seaman A.Knapp on the crop

(*a*) Paddy (*b*) Maize
(*c*) Cotton (*d*) Sugarcane

77. A farmer brings infected ears of *bajra* to the Subject Matter Specialist (SMS) for solution of the disease.The ear of bajra here is example of

(*a*) Model (*b*) Mock-up
(*c*) Illustration (*d*) Specimen

78. To prove the worth of new practice suited to a particular area, the best method of extension education is

(*a*) Method Demonstration (*b*) Field travel
(*c*) Farm visit (*d*) Result demonstration

79. Extension workers are

(*a*) Professional leaders (*b*) Autocratic leaders
(*c*) Functional leaders (*d*) Caste leaders

80. In India 20 point programme was launched by

(*a*) Manmohan Singh (*b*) Rajiv Gandhi
(*c*) Indira Gandhi (*d*) Mahatma Gandhi

81. The originator and leader of Farmers Cooperative Demonstration movement was

(*a*) A.B. Graham (*b*) I.P. Robers
(*c*) J.S. Woodward (*d*) Seaman A. Knapp

82. General meeting are broadly,the meeting of

(*a*) Heterogeneous participation (*b*) Homogeneous participation
(*c*) Community participation (*d*) None of these

83. Rural development depends on

(*a*) Research (*b*) Research - Extension
(*c*) Research - Teaching -Extension (*d*) Only Teaching

84. Campaign is an

(*a*) Intensive teaching (*b*) Extensive teaching
(*c*) Invasive teaching (*d*) None of these

85. Extension education is

(a) Formal education
(b) Informal education
(c) Non-formal education
(d) None of these

86. The father of demonstration in extension is

(a) S.A. Knapp
(b) Rober Chambers
(c) J.P. Leagans
(d) August Comte

87. The major drawback of mass media in context of extension work is

(a) Expensive
(b) Not easily available
(c) Lack of free feedback
(d) None of these

88. Seaman A Knapp conducted result demonstration in

(a) Maize
(b) Cotton
(c) Barley
(d) Rice

89. The main function of agricultural universities in India is

(a) Technology generation and dissemination
(b) Research, Training and Extension
(c) Research, Teaching and Extension
(d) None of these

90. An agricultural innovator is a

(a) Traditional Leader
(b) Functional Leader
(c) Opinion Leader
(d) All of these

91. Adoption is basically a

(a) Social process
(b) Mental process
(c) Psychological process
(d) None of these

92. Poster is most effective at which stage

(a) Evaluation
(b) Awareness
(c) Testing
(d) Reporting

93. The adopter categories who is also considered as the custodians of indigenous knowledge

(a) Innovators
(b) Laggards
(c) Early adopters
(d) None of these

94. 'Legitimization' is done by

(a) Late majority
(b) Early majority
(c) Early adopters
(d) Late adopters

95. In USA extension service is called

(a) National Extension Service
(b) Cooperative Extension Service
(c) Coordinating Extension Service
(d) None of these

96. Major components of human behaviour are

(a) Knowledge (b) Skill
(c) Attitude (d) All of these

97. The most basic social institution is

(a) Village (b) Family
(c) Class (d) District

98. Learning is motivated by

(a) Self (b) Group
(c) Environment (d) Family

99. Phillip 66 Format is also known as

(a) Brain storming (b) Forum
(c) Buzz session (d) None of these

100. The type of leader who has received specialized training and is paid for the field work is called

(a) Autocratic leader (b) Democratic leader
(c) Professional leader (d) None of these

101. Dr. Seaman A.Knapp was professor of agriculture at

(a) Union College, Schenectady (b) Royal College, Massachusetts
(c) Agriculture College, IIlions (d) State College, Iowa

102. In USA the extension work is named

(a) Federal extension work (b) Federal extension service
(c) Cooperative extension service (d) Cooperative extension work

103. In USA, National Board on Agriculture was established in

(a) 1799 (b) 1796
(c) 1790 (d) 1786

104. The first ever journal of extension services was published from

(a) India (b) Germany
(c) USA (d) UK

105. National 4-H club center in USA is located

(*a*) Iowa (*b*) New York
(*c*) Philadelphia (*d*) Washington

106. International Farm Youth Exchange (IFYE) programme began in the year

(*a*) 1938 (*b*) 1948
(*c*) 1958 (*d*) 1968

107. The term 'Extension' has originated in

(*a*) Persian (*b*) English
(*c*) Latin (*d*) Greek

108. Extension literally means

(*a*) Talking with (*b*) Thinking about
(*c*) Deciding for (*d*) Stretching out

109. Extension can be considered as a

(*a*) Service (*b*) Profession
(*c*) Discipline (*d*) All the these

110. The term university extension was first used in

(*a*) 1946 (*b*) 1972
(*c*) 1950 (*d*) 1921

111. The term extension was originated in

(*a*) Latin (*b*) Persian
(*c*) Greek (*d*) English

112. Extension basically is two way process between

(*a*) Researcher and planner (*b*) Farmer and researcher
(*c*) Teacher and planner (*d*) Researcher and planner

113. An out-school system of education for rural people

(*a*) Management (*b*) Participation
(*c*) Communication (*d*) Extension

114. "Challenging the Professionals " is a book written by

(*a*) Paul Daniel (*b*) N. Roling
(*c*) G.L. Ray (*d*) R. Chambers

115. Which is basic unit of extension work ?

(*a*) Individual (*b*) Community

(*c*) Society (*d*) Family

116. An extension worker is a

(*a*) Voluntary leader (*b*) Professional leader

(*c*) Lay leader (*d*) All of these

117. Method demonstration works on the principle of

(*a*) Learning is believing (*b*) Learning by teaching

(*c*) Learning by doing (*d*) All of these

118. Result demonstration works on the principle of

(*a*) Learning is believing (*b*) Learning by teaching

(*c*) Learning by doing (*d*) Seeing is believing

119. The terminology that does to relates with the most innovative member of a social system

(*a*) Ultra adopters (*b*) Drones

(*c*) Progressists (*d*) None of these

120. The percentage of population constituted by the innovators

(*a*) 10.7 per cent (*b*) 16.5 per cent

(*c*) 4.5 per cent (*d*) 2.5 per cent

121. The character best represent an innovator

(*a*) Skeptical (*b*) Venturesome

(*c*) Traditional (*d*) All of these

122. The extension work must be based on

(*a*) Need and interests of the people (*b*) Interests of local leaders

(*c*) Interests of BDO (*d*) Interest of teacher

123. The fundamental objective of extension education is

(*a*) Modernization (*b*) Development of Pradhans

(*c*) Development of traders (*d*) None of these

124. The book "Extension Education" is written by

(*a*) Adivi Reddy (*b*) A.A. Reddy

(*c*) O.P. Dahama (*d*) G.L. Ray

125. Diffusion has

(a) Five elements (b) Four elements
(c) Six elements (d) Ninth elements

126. Proportion of laggards in farming community is

(a) 16 per cent (b) 25 per cent
(c) 30 per cent (d) 100 per cent

127. The name 4-H to the working youth clubs in USA was given by

(a) Seaman A. Knapp (b) A.B. Graham
(c) O.H. Benson (d) A.B. Ross

128. The motto of 4-H club was proposed by

(a) B.C. Bailey (b) L.H. Bailey
(c) W.C. Stallings (d) Carri Harrison

129. The 4-H pledge was worded by

(a) I.P. Roberts (b) Carri Harrison
(c) W.A. Henry (d) Otis Hall

130. The 4-H club motto is

(a) To make the best better (b) To make the better best
(c) To make the good better (d) To make the good best

131. Education cause

(a) Change in Knowledge (b) Change in attitude
(c) Change in skills (d) All of these

132. The Latin roots for the word' extension'

(a) ex and tension (b) ex and tensio
(c) ex and luceo (d) None of these

133. 'Vulgarisation; in French is equivalent to the word

(a) Social-taboos (b) Social change
(c) Education (d) Extension

134. The Dutch word for extension " Voorlichting" means

(a) Advisory work (b) Improving skills
(c) Simplification (d) Lighting the path

135. The book entitled "Extension Science" is written by

(a) Theodore Lips (b) N. Roling
(c) G.L. Ray (d) Edgar Dale

136. Method demonstration as an extension teaching method is classified under

(*a*) Individual contact (*b*) Mass contact

(*c*) Group contact (*d*) Both (*a*) and (*b*)

137. There are ______ levels of objectives in extension

(*a*) Six (*b*) Four

(*c*) Five (*d*) Three

138. The process of using a new idea on continuous basis is

(*a*) Diffusion (*b*) Utility

(*c*) Adoption (*d*) Innovation

139. The person who adopts new practices at the last is

(*a*) Early adopter (*b*) Innovator

(*c*) Laggard (*d*) Late adopter

140. The cone of experience as given by Dale was revised by

(*a*) Leagans (*b*) Sheal

(*c*) Wilkening (*d*) W. N. Newman

141. Name of the person who is regarded as father of extension in USA

(*a*) Dr. Seaman A. Knapp (*b*) Dr. Robert Chambers

(*c*) Dr. Augustone Frude (*d*) S.K. Day

142. A small farmer is one who own land upto

(*a*) 1 hectare (*b*) 2 hectares

(*c*) 8 hectares (*d*) 4 hectares

143. The cone of experience was revised by Sheal in the year

(*a*) 1976 (*b*) 1980

(*c*) 1985 (*d*) 1989

144. The last stage in the extension education process is

(*a*) Evaluation (*b*) Reconsideration

(*c*) Adoption (*d*) All of these

145. The fundamental objective of extension work is

(*a*) Development of farmer (*b*) Development of individual

(*c*) Development of Culture (*d*) None of these

146 Farmers first model was proposed by

(*a*) G.H. Mead (*b*) R. Chambers
(*c*) Van den ban (*d*) G.W. Weision

147. The term extension education was first coined in

(*a*) China (*b*) UK
(*c*) Netherlands (*d*) India

148. The basic unit of extension work is

(*a*) Individual (*b*) Society
(*c*) Family (*d*) None of these

149. Extension can be effective only through

(*a*) Democratic change (*b*) Voluntary change
(*c*) Autocratic change (*d*) Persuasive change

150. In which year Agricultural Extension Service was started in USA?

(*a*) 1862 (*b*) 1890
(*c*) 1940 (*d*) None of these

151. In the following which is not a goal of agricultural extension?

(*a*) Change in skill (*b*) Change in religion
(*c*) Change in attitude (*d*) None of these

152. Puppet (Kathputli) is

(*a*) Audio-aid (*b*) Visual aid
(*c*) Both (*a*) and (*b*) (*d*) None of these

153. In India, first time T.V. was used as mass media in the year

(*a*) 1975 (*b*) 1960
(*c*) 1967 (*d*) 1970

154. The National Demonstration Scheme was launched in the year

(*a*) 1975 (*b*) 1960
(*c*) 1970 (*d*) 1965

155. What is the principles of extension education?

(*a*) Learning by reading
(*b*) Learning by doing
(*c*) Learning by seeing
(*d*) Learning by hearing

156. Lab to land programme was started by

(a) ICAR
(b) Govt. of India
(c) UPCAR
(d) None of these

157. CAPART refers to

(a) Centre for Advancement of People's Action and Rural Technology.
(b) Centre for Agriculture, People's Action and Rural Technology.
(c) Council for Advancement of People's Action and Rural Technology.
(d) Council for Agriculture, People's Action and Rural Development.

158. Help Line Service refers to

(a) To provide Agricultural Information to farmers over telephone.
(b) To provide Agricultural Loan to farmers.
(c) To provide Agricultural inputs to farmers.
(d) All of these

159. ATIC refers to

(a) Agricultural Technological information Centre
(b) Agricultural Technology Information Centre
(c) Advance Technology of Information Centre
(d) None of these

160. Training and Visit System of Agricultural Extension was propounded by

(a) Daniel M.D. Beonar
(b) Kalim Qamar
(c) Van Den Ban
(d) Axinn

FILL IN THE BLANKS

1. The term 'Extension Education' was first coined in ___________
2. The father of extension in India is _________
3. The father of extension education is________
4. Cone of experience was given by________
5. The 4H club was founded by _________
6. The research by Ryan and Cross was on _________
7. The book "Extension Science" is written by_________
8. The basic unit of extension work is ________

9. A small farmer is one who own land upto________
10. The Indian Council of Agricultural Research established as section of Extension Education at headquarters in the year______
11. The 4-H pledge was worded by _______
12. the Motto of 4-H club was proposed by______
13. In USA the extension work is named as_______
14. Hoshangabad model of extension emphasis on___________
15. TRYSEM was launched in the year________
16. Kisan Call Centre (KCC) was launched in the year________
17. Training and Visit System of agricultural extension was propounded by___
18. MANAGE Extension Review is from________
19. The Chairman of NITI Aayog is________
20. India Lives in its Village said by____________

MARK TRUE OR FALSE

1. In 1957 Ensminger said extension is education.
2. In India, 20 point programme was launched by Rajiv Gandhi.
3. The National Demonstration Scheme was launched in the year 1965.
4. Principle of extension education is learning by hearing.
5. Lab to land programme was started by UPCAR.
6. Helpline service refers to provide agricultural information to farmers over telephone.
7. ATIC refers to Agricultural Technology Information Centre.
8. The word extension first used by Edgar Dale.
9. The concept of bio-village was conceived by Dr. M.S. Swaminathan.
10. August Comte is considered as father of Anthropology.
11. S.K. Dey was associated with Community Development Programme.
12. Anand model for rural development focuses on milk.
13. Private Extension is mostly paid extension.
14. CAPART refers to Centre for Advancement of People's Action and Rural Technology.
15. Extension workers are professional leaders.
16. The originator and leader of Farmers Co-operative Demonstration Movement was Seaman A. Knapp.
17. County is the unit of extension work is China.

18. Phillip 66 format is also known as Buzz session
19. Campaign is an intensive teaching
20. The Dutch word for extension 'Voorlichting" means lighting the path

ANSWERS

Multiple Choice Questions

1.(b)	2.(c)	3.(a)	4.(a)	5.(d)	6.(a)	7.(a)
8.(a)	9.(d)	10.(b)	11.(a)	12.(a)	13.(a)	14.(d)
15.(b)	16.(b)	17.(c)	18.(c)	19.(b)	20.(b)	21.(b)
22.(c)	23.(a)	24.(d)	25.(a)	26.(b)	27.(a)	28.(b)
29.(a)	30.(d)	31.(c)	32.(c)	33.(a)	34.(c)	35.(b)
36.(a)	37.(d)	38.(b)	39.(d)	40.(a)	41.(a)	42.(c)
43.(d)	44.(c)	45.(b)	46.(c)	47.(b)	48.(c)	49.(b)
50.(c)	51.(d)	52.(d)	53.(a)	54.(a)	55.(a)	56.(d)
57.(c)	58.(b)	59.(b)	60.(a)	61.(c)	62.(d)	63.(a)
64.(c)	65.(b)	66.(c)	67.(d)	68.(a)	69.(d)	70.(c)
71.(b)	72.(d)	73.(d)	74.(b)	75.(b)	76.(c)	77.(d)
78.(d)	79.(a)	80.(c)	81.(d)	82.(a)	83.(c)	84.(a)
85.(c)	86.(a)	87.(c)	88.(b)	89.(c)	90.(b)	91.(b)
92.(b)	93.(b)	94.(c)	95.(b)	96.(d)	97.(b)	98.(a)
99.(c)	100.(c)	101.(d)	102.(c)	103.(b)	104.(c)	105.(d)
106.(b)	107.(c)	108.(d)	109.(d)	110.(d)	111.(a)	112.(b)
113.(d)	114.(d)	115.(d)	116.(b)	117.(c)	118.(d)	119.(b)
120.(d)	121.(b)	122.(a)	123.(a)	124.(b)	125.(c)	126.(a)
127.(c)	128.(d)	129.(d)	130.(a)	131.(d)	132.(b)	133.(d)
134.(d)	135.(d)	136.(c)	137.(d)	138.(c)	139.(c)	140.(b)
141.(a)	142.(b)	143.(d)	144.(b)	145.(b)	146.(b)	147.(b)
148.(c)	149.(b)	150.(a)	151.(b)	152.(c)	153.(c)	154.(d)
155.(d)	156.(a)	157.(c)	158 (a)	159.(b)	160.(a)	

Fill in the Blanks

1. UK
2. K.N. Singh
3. J.P. Leagons
4. Edger Dale
5. O.H. Benson
6. Hybrid Corn
7. N. Roling
8. Family
9. 2 hectares
10. 1971
11. Otis Hall
12. Carri Harrison
13. Co-operative Extension Services
14. Public-Pvt. Partnership
15. 1979
16. 2004
17. M.D. Beonar
18. Hyderabad
19. Prime Minister
20. Mahatma Gandhi

Mark True or False

1. True
2. False
3. True
4. True
5. False
6. True
7. True
8. False
9. True
10. False
11. True
12. True
13. True
14. False
15. True
16. True
17. False
18. True
19. True
20. True

CHAPTER 2

AGRICULTURAL COMMUNICATION, MEDIA AND INFORMATION TECHNOLOGY

Multiple Choice Questions

1. **Which is the most basic fumigation of communication ?**
   (*a*) Integration (*b*) Influence
   (*c*) Instruction (*d*) Information
2. **Round Table discussion is known as**
   (*a*) Symposium (*b*) Workshop
   (*c*) Panel (*d*) Forum
3. **Sponsoring agency of Cyber Extension project is**
   (*a*) NAARM, Hyderabad (*b*) GBPUA&T, Pantnagar
   (*c*) MANAGE, Hyderabad (*d*) ICAR, New Delhi
4. **Sponsoring agency of Agricultural Gateway of India is:**
   (*a*) NAARM, Hyderabad (*b*) MANAGE, Hyderabad
   (*c*) ICAR, New Delhi (*d*) None of these
5. **e-chaupal launched by ITC in :**
   (*a*) July, 2000 (*b*) June, 2000
   (*c*) July, 1988 (*d*) June, 1977
6. **ICT-Project 'Dairy Cooperative' developed by**
   (*a*) NDDB (*b*) PARAG
   (*c*) AMULYA (*d*) NDRI

7. **Transfer of technology is related with**
   (a) Progressive idea (b) Local idea
   (c) Foreign idea (d) Both (b) and (c)

8. **Theory of human behaviour in diffusion process is given by**
   (a) Parsons (b) Aristotle
   (c) Rogers (d) Cuber

9. **The disequilibrium theory was given by**
   (a) MacIver (b) Lewin
   (c) Rogers (d) Haver

10. **CONA stands for**
    (a) Community Oriented Need Assessment
    (b) Career Oriented Need Analysis
    (c) Creative Organizational Need Assessment
    (d) Career Oriented Need Assessment

11. **DNS in Internet technology stands for**
    (a) Domain Name System (b) Dynamic Name System
    (c) Distributed Name System (d) Doppler Name System

12. **HTML stands for**
    (a) Hyper Text Manipulating Links
    (b) Hyper Text Manipulation Language
    (c) Hyper Text Managing Links
    (d) Hyper Text Markup Language

13. **Communication with oneself is known as**
    (a) Organisational Communication
    (b) Interpersonal Communication
    (c) Intrapersonal Communication
    (d) None of these

14. **The term 'SITE' stands for**
    (a) Satellite Indian Teachers Experiment
    (b) Satellite International Television Experiment
    (c) Satellite Instructional Teachers education
    (d) Satellite Instructional Television Experiment

15. MC National University of Journalism and Communication is situated at

(a) Bhopal (b) Hyderabad

(c) Jodhpur (d) Mumbai

16. Communication is sharing together and is for

(a) Popularity (b) Differences

(c) Apathy (d) Commonness

17. Communication I.Q. was studies by

(a) Albert Mayer (b) F. L. Bryne

(c) Bhanja (d) Hovland

18. Which of the following is a traditional and simple aid communication

(a) Television (b) Radio

(c) Model (d) Puppet

19. The communication skills that get emphasis on listening and reading

(a) Encoding (b) Effect

(c) Attitude (d) Decoding

20. Speaking and writing involve which type of communication skills

(a) Knowledge (b) Feedback

(c) Decoding (d) Encoding

21. The abbreviation COLK in COLK fallacy" means

(a) Clear Only if Know

(b) Clarity of Informational Knowledge

(c) Capacity of Indigenous Know-how

(d) Concept of Imaging Knowledge

22. Letters for communication in extension should follow the "4'S Formula". "4'S include

(a) Shortness, Simplicity, Strength, Sincerity

(b) Swiftness, Shortness, Strength, simplicity

(c) Steadfast, Swiftness, Strength, Shortness

(d) Severity, Shortness, Swiftness, Sincerity

23. The percentage of learning that takes place through the sense of hearing is

(a) 1.0 per cent (b) 1.5 per cent

(c) 7.0 per cent (d) 3.5 per cent

24. The appropriate height of letters of a demonstration poster should be

(a) 0.5 inches
(b) 1.0 inches
(c) 1.5 inches
(d) 2.5 inches

25. Which one of the following element of communication ?

(a) Source, message, channel, receiver and feedback
(b) Trainer, channel, learner and feedback
(c) Message, channel, learner and feedback
(d) Teacher, message, receiver and feed back

26. ABC of Journalism is

(a) Accurate, Brevity, Clear
(b) Accuracy, Brief, Complete
(c) Accuracy, Brevity, Clarity
(d) None of these

27. Apple satellite launched in the year

(a) 1971
(b) 1981
(c) 1989
(d) 1979

28. If there is a problem of plague in a particular area, the effective method used to control the disease is

(a) Awareness Campaign
(b) Awareness through Radio
(c) Awareness Television
(d) Awareness through Exhibition

29. The screen recommended for viewing a three dimensional image is

(a) Translucent
(b) Aluminum
(c) Plastic Screen
(d) Glass screen

30. Video-Conferencing can be classified as one of the following types of communication

(a) Visual one way
(b) Visual two way
(c) Audio -Visual one way
(d) Audio -Visual two way

31. A series of cards presented before an audience one by one to tell a complete story is called

(a) Flash cards
(b) Flannel graph
(c) Pie Chart
(d) Bar Graph

32. A series of still pictures on one roll is known as

(a) Channel
(b) Film strip
(c) Mock -Up
(d) Flash card

33. A bulletin should contain number of pages

(*a*) 8-12 (*b*) 12 - 24

(*c*) 24 - 48 (*d*) 12-18

34. In India, slightly less than half the village are inhabited by ____ persons

(*a*) < 500 (*b*) < 900

(*c*) <2000 (*d*) <1500

35. About 3/4 Indian villages have a human population of less than

(*a*) 400 (*b*) 800

(*c*) 2000 (*d*) 1000

36. 'Communication is the first principle in the society' are the words of

(*a*) Woodraw W. Sayre (*b*) W.N. Newman

(*c*) Peter Drucker (*d*) Edgar Dale

37. In India, satellite experiment was started in the year

(*a*) 1965 (*b*) 1985

(*c*) 1975 (*d*) 1970

38. Communication involves three phases expression, interpretation and

(*a*) Response (*b*) Re-selection

(*c*) Transmission (*d*) None of these

39. A five minute radio talk has

(*a*) 500 words (*b*) 200 words

(*c*) 600 words (*d*) 700 words

40. The problem concerned with the accuracy of the transference of information from sender to receiver comes under

(*a*) Technical problems (*b*) Semantic problems

(*c*) Psychological problems (*d*) Non-technical problems

41. The 4-is of communication are

(*a*) Informal, Instruction, Influence and Integration

(*b*) Intermediation, Information, Interest and Instruction

(*c*) Information, Intention, Influence and Instruction

(*d*) Information, Instruction, Influence and Integration

42. The communication, which takes place within an individual, is known as

(a) Intrapersonal communication

(b) Interpersonal communication

(c) Internal communication

(d) None of these

43. "Communication I.Q. test" was given by

(a) Paul Deweny (b) Paul Herbert

(c) Paul Preston (d) None of these

44. Agricultural programme "Krishi Darshan" was firstly started in the year

(a) 1957 (b) 1967

(c) 1960 (d) 1970

45. Television was separated from Radio by the Ministry of Broadcasting on

(a) 1st April 1976 (b) 1st April 1970

(c) 1st August 1975 (d) 1st August 1978

46. "Extension digest" is published from

(a) MANAGE, Hyderabad (b) IARI, New Delhi

(c) ICAR, New Delhi (d) GBPUA&T, Pantnagar

47. Indirect and International feedback is available in

(a) Mass Media (b) Visual Media

(c) Audio Media (d) Group Media

48. Soyachoupal is more beneficial mainly for

(a) Large holding farmers (b) Medium holding farmers

(c) Small holding farmers (d) None of these

49. e-chaupal has

(a) WLL connectivity (b) VSAT connectivity

(c) LAN connectivity (d) WAN connectivity

50. COW has

(a) GPRS connectivity (b) LAN connectivity

(c) WAN connectivity (d) VSAT connectivity

51. Unstrumented channels of communication are:

(a) Talk in the field and on the road (b) Talk at home and at the well

(c) Talk in tea hour and coffee hours (d) All of these

52. Which is the pioneering institution to draw upon folk music?

(*a*) Doordarshan
(*b*) All India Radios
(*c*) Akaswani
(*d*) Star Plus

53. Folk arts laying a meaningful role in rural areas in educating the

(*a*) Student
(*b*) Tribal people
(*c*) Urban people
(*d*) Rural people

54. "Hue" indicates

(*a*) Brightness of colour
(*b*) Name of colour
(*c*) Quality of colour
(*d*) Darkness of colour

55. National Documentation Center on Mass Communication was created on the year

(*a*) 1976
(*b*) 1978
(*c*) 1985
(*d*) 1992

56. Indian Institute of Mass Communication is situated at

(*a*) Lucknow
(*b*) New Delhi
(*c*) Mumbai
(*d*) Tamil Nadu

57. Communication channels are the physical bridges between sender and receiver of the message" are the words of

(*a*) Kelsey
(*b*) Herzberg
(*c*) Leagans
(*d*) Hearnel

58. The "COLK fallacy" is associated with

(*a*) TV channels
(*b*) Audio-Visual aids
(*c*) Radio
(*d*) Transfer of technology

59. The concept of "COLK fallacy" has been proposed by

(*a*) J.P. Leagons
(*b*) Alexander Kapp
(*c*) Edgar Dale
(*d*) H.W. Butt

60. Headquarter of Indian Society of Extension Education (ISEE) is located at

(*a*) New Delhi
(*b*) Hyderabad
(*c*) Jabalpur
(*d*) Pantnager

61. "Journal of Extension Education" is published from

(*a*) NDUAT, Kumarganj
(*b*) ICAR, New Delhi
(*c*) TNAU, Coimbatore
(*d*) GBPUA&T, Pantnagar

62. "Indian Research Journal of Extension Education" is published from

(a) New Delhi *(b)* Hyderabad

(c) Agra *(d)* Lucknow

63. In India, Television broadcast for rural development started in

(a) 1950 *(b)* 1957

(c) 1960 *(d)* 1984

64. Radio mass medium is characteristic by

(a) One way without instant audience response

(b) One way with immediate feedback

(c) one way and colourful

(d) None of these

65. In the following, which is not a method of group communication?

(a) Demonstration *(b)* Conference

(c) Flannel graph *(d)* Circular letter

66. Communication is the process by which message are transferred from source to

(a) Source *(b)* Channel

(c) Receiver *(d)* all of these

67. Leaflet is a single sheet of paper folded to make a ____ page piece of printed matter.

(a) 2 *(a)* 8

(c) 12 *(d)* 4

68. For one talk, how many flash cards should be used

(a) 10 -12 *(a)* 8 - 10

(c) 16 - 20 *(d)* 21 - 24

69. The process of effecting on interchange of understanding between two or more people is known as

(a) Diffusion *(b)* Innovation

(c) Communication *(d)* None of these

70. Traditional media may compute for attention will modern media such as

(a) TV *(b)* Radio

(c) Cassettes or Video *(d)* All of these

71. Traditional channel of communication (Munadi) is more effective than______by KVK worker and newspaper in disseminating message of local importance in the tribal villages

(*a*) Official contact
(*b*) Intrapersonal contact
(*c*) Interpersonal contact
(*d*) None of these

72. Which one of the following bank used puppetry show in Pantnagar farmers fair to arouse the interest of the visitors in bank savings and loans?

(*a*) Bank of India
(*b*) Punjab National Bank
(*c*) Central Bank
(*d*) State Bank of India

73. Communicator is the characteristic element in the model of

(*a*) Lazwell
(*b*) Rogers
(*c*) Durkheim
(*d*) Leagons

74. News paper article are included in which method of extension

(*a*) Individual
(*b*) Group
(*c*) Mass
(*d*) Both (*a*) and (*b*)

75. Which is the less intensive and less effective method of extension?

(*a*) Individual
(*b*) Group
(*c*) Mass
(*d*) None of these

76. Flash cards, pull chart, slides and film strips are the type of aids

(*a*) Audio
(*b*) Visual
(*c*) Audio - Visual
(*d*) None of these

77. A working model is known as

(*a*) Objects
(*b*) Model
(*c*) Mock-up
(*d*) Poster

78. Real object taken out of their natural setting is known as

(*a*) Specimens
(*b*) Objects
(*c*) Model
(*d*) Mock -up

79. The communication strategy embedded in the transfer of technology paradigm is

(*a*) Persuasive +Participatory
(*b*) Persuasive+ Paternalistic
(*c*) Educational + Participatory
(*d*) Educational + Paternalistic

80. The English equivalent of the word *'Communis'* is

(a) Information | (b) **Common**
(c) Talk | (d) None of the above

81. The word communication have originated from

(a) Roman | (b) Latin
(c) Greek | (d) English

82. Trustworthiness and competency are element of

(a) Empathy | (b) Fidelity
(c) Credibility | (d) None of the above

83. Reading newspaper is

(a) Interpersonal communication | (b) Mass communication
(c) Intrapersonal communication | (d) Dynamic communication

84. The number of elements in the communication model of Berlo

(a) 7 elements | (b) 6 elements
(c) 5 elements | (d) None of these

85. The Leagon's model of communication contains

(a) 7 elements | (b) 6 elements
(c) 5 elements | (d) None of these

86. The S-M-C-R model of communication is proposed by

(a) Thurston | (b) Leagons
(c) Berlo | (d) George

87. Destination is an important element in the model of

(a) Thorner | (b) Schramm
(c) Berlo | (d) None of these

88. Source-transmitter-signal-receiver-destination is the model of communication is given by

(a) Edgar Dale | (b) Leagon, J. Paul
(c) Shannon and Weaver | (d) Westley and Machean

89. While reading any visual, human eye moves normally in a

(a) 'Z' like pattern | (b) 'Y' like pattern
(c) 'W' like pattern | (d) 'X' like pattern

90. The most appropriate letter size for titles in a chart is

(a) 1.5 inch (b) 3.5 inch

(c) 2.5 inch (d) 4.0 inch

91. The size of flash cards for a group of 30 to 50 persons is

(a) 20 x 30 inches (b) 20 x 25 inches

(c) 15 x 20 inches (d) 10 x 12 inches

92. The 'COIK' fallacy is associated with

(a) Communication channels (b) Communicator

(c) Audiovisual aids (d) Receiver

93. The ratio maintained between 'f' number and shutter speed of a camera is

(a) 3:1 (b) 2:1

(c) 1:3 (d) 1:2

94. Bunch of loose printed papers properly folded offered for sale at regular intervals

(a) Newspaper (b) Newsletter

(c) Annual (d) Journal

95. The book titled " Writing for Farm Families" is written by

(a) Van den Ban (b) Richard Dance

(c) MG Kamath (d) R Chambers

96. The basic colour type used in printing

(a) Red, Blue, Green (b) Cyan, Magenta, Red

(c) Red, Blue, Yellow (d) Red, Blue, Orange

97. The first farm magazine in India is

(a) Phool Phul (b) Kheti

(c) Krishi Chayanika (d) Yojana

98. The Printing ink consists of

(a) Lead, Zinc and Antimony (b) Tin, Zinc and Lead

(c) Lead, Zinc and Tin (d) Lead, Tin, and Antimony

99. Bengal Gazette was published from

(a) Mumbai (b) Calcutta

(c) Chennai (d) Lucknow

100. Prasar Bharti came into existence on

(a) 23rd January 1997 (b) 18th January 1996

(c) 15th October 1995 (d) 20th December 1996

101. Radio broadcasting in India started from

(a) Delhi (b) Mumbai

(c) Calcutta (d) Hyderabad

102. The effective rate of delivery for a radio talk is

(a) 210-240 words/minute (b) 150-170 words/minute

(c) 120-140 words/minute (d) 60-110 words/minute

103. The first telecast on Television in India was started on

(a) 15th November 1969 (b) 15th October 1959

(c) 15th September 1959 (d) 15th August 1960

104. Who is the author of book " The process of communication book is written by ?

(a) David K. Berlo (b) J.P. Leagons

(c) Paul Herbert (d) Kelsey and Hearne

105. The mathematical theory of communication is proposed by

(a) E. Dale (b) D. Esminger

(c) A. Stevens (d) C. Shannon and W. Weaver

106. " Extension digest " is published from

(a) MANAGE (b) IARI

(c) GBPUAT (d) NDRI

107. Effective communication needs a supportive

(a) Economic Environment (b) Social Environment

(c) Political Environment (d) None of these

108. A major barrier in the transmission of cognitive data in the process of communication is an individual's

(a) Expectation (b) Coding ability

(c) Social status (d) None of these

109. Each character on the keyboard of computer has an ASCII value which stands for

(a) American Standard Code for Information Interchange

(b) American Stock Code for Information Change

(c) African Standard Code for Information Change

(d) None of these

110. TCP/IP is necessary if one is to connect the

(a) Internet (b) Phone lines

(c) LAN (d) None of these

111. The term "TRP" is associated with TV shows stands for

(a) Total Rating Points (b) Term Rating Points

(c) Thematic Rating Points (d) Television Rating Points

112. The term 'DAVP' stands for

(a) Directorate of Advertising and Vocal Publicity

(b) Division of Audio-Visual Publicity

(c) Directorate of Advertising and Visual Publicity

(d) None of these

113. The movie film recording is made by

(a) Magnetic (b) Mechanical

(c) Optical (d) Chemical

114. Visual teaching aid flannel graph is also called as

(a) Chalk board (b) Bulletin board

(c) Khadder graph (d) Black board

115. Puppet show as a method of transfer of technology is an example of

(a) Individual media (b) Group media

(c) Mass media (d) None of these

116. The main mode of extension in the farmer first model

(a) Researcher to farmer (b) Agent to farmer

(c) Researcher to agent (d) Farmer to farmer

117. Farmers first model was given by

(a) G.L. Ray (b) R M Rogers

(c) R. Chambers (d) J.P. Leagons

118. Educational television is classified under

(a) Broadcast media (b) Screen media

(c) Printed media (d) Both (b) and (c)

119. Which company is providing mobile service with name 'Cell One' to the consumers ?

(a) BSNL (b) Vodafone

(c) Idea (d) MTNL

120. The main problem with mass communication is

(*a*) Costly (*b*) Same information

(*c*) No local dialect (*d*) Cumbersome

121. Cybernetics is the science of

(*a*) Decision Making (*b*) Organization

(*c*) Communication (*d*) Environment

122. The Indian Institute of Mass Communication is situated at

(*a*) New Delhi (*b*) Chandigarh

(*c*) Mumbai (*d*) Nagpur

123. Communication breakdown is a common feature of

(*a*) Ulterior transaction (*b*) Straight transaction

(*c*) Vertical transaction (*d*) Cross transaction

124. Balance theory of communication is given by

(*a*) Heider (*b*) Cronkhite

(*c*) Axinn (*d*) S.K. Dey

125. 'Communicator' is the characteristic element in the model of

(*a*) Lazwell (*b*) Leagons

(*c*) Osgood (*d*) Gullick

126. The best screen for a well-lighted room is

(*a*) Traditional (*b*) Lenticular

(*c*) Matte (*d*) Plastic

127. Colour with highest value is

(*a*) Red (*b*) Green

(*c*) Yellow (*d*) White

128. Black and White colours are

(*a*) Achromatic colours (*b*) Chromatic colours

(*c*) Diligent colours (*d*) Symbolic colours

129. The number of pages a leaflet contain

(*a*) Seven (*b*) Twelve

(*c*) Four (*d*) Eight

130. The optimum number of flash cards is

(a) 8-12 (b) 16-20
(c) 12-14 (d) 10-12

131. The ABC of journalism is related to

(a) Accuracy, Brevity, Clarity (b) Accuracy, Brevity, Credibility
(c) Accountable, Brief, Clear (d) Active, Brief, Clear

132. Which one is not an element of communication process?

(a) Message (b) Fidelity
(c) Channel (d) None of these

133. Primary colour or basic colour set is

(a) Red, Blue, Yellow (b) Red, White, Grey
(c) Red, Purple, Green (d) Red, Sky blue, Lemon

134. According to hypodermic needle model of communication, mass media has

(a) Direct effect on mass audience
(b) Indirect effect on mass audience
(c) Secondary effect on mass audience
(d) Tertiary effect on mass audience

135. Which type of market plays an important role in the life of the tribal people is

(a) Monthly market (b) Daily market
(c) Seasonal market (d) Weekly market

136. The tribal market brings together people from different ethnic group for not only economic in the tribal region but also

(a) Secular activities (b) Secular and religious activities
(c) Religious activities (d) None of these

137. The market is the most powerful channel of communication in the

(a) Urban region (b) Tribal region
(c) Rural region (d) None of these

138. The impact of weekly market on their traditional life has shown an attitude of accepting

(a) Communication (b) Diffusion
(c) Innovation (d) All of these

139. Teaching act as

(*a*) Catalyst (*b*) Analyst

(*c*) Both (*a*) and (*b*) (*d*) All of these

140. Which is the process of communication ?

(*a*) Learning (*b*) Teaching

(*c*) Both (*a*) and (*b*) (*d*) None of these

141. Which of the following is the traditional media of communication is ?

(*a*) Slide projector (*b*) Radio

(*c*) Television (*d*) Ram Leela

142. In which year Indian television started broadcasting of rural programmes?

(*a*) 1956 (*b*) 1960

(*c*) 1971 (*d*) 1967

143. Action-reaction interdependence in communication is known as

(*a*) Credibility (*b*) Fidelity

(*c*) Message treatment (*d*) Feedback

144. While communicating verbally how much message is distorted?

(*a*) 30 per cent (*b*) 70 per cent

(*c*) 60 per cent (*d*) 80 per cent

145. "Krishi Darshan programme in India is telecasted from

(*a*) Hyderabad (*b*) Mumbai

(*c*) Chennai (*d*) New Delhi

146. ABC of poster stands for

(*a*) Attractive, Brief and Clear (*b*) Attractive, Brief and Complete

(*c*) Attractive, Brief and Creative (*d*) Attractive, Brief and Clean

147. "The communication process in rural development" is written by

(*a*) J.P. Leagans (*b*) Paul Stephenson

(*c*) Max Weber (*d*) August Comte

148. Leagons model of communication process is

(*a*) Communication -Message-Channel-Treatment- Audience-Audience response

(*b*) Sender - Encoding- Channel - Decoding – Receiver

(*c*) Source- Encoder - Message Channel-Decoder-Communication receiver

(*d*) Communication-Encoder-channers-Receivers

149. An electronic audio- visual medium which provide pictures with synchronized sound is

(a) Television (b) Radio

(c) Films (d) Telephone

150. The term electronic chalkboard refers to

(a) LCD projector (b) VHS projector

(c) OHP projector (d) Beta chrome projector

151. The ideal letter size for a poster to be displayed in a hall of 25 m length

(a) 1.4 inch (b) 2.1 inch

(c) 2.7 inch (d) 3.5 inch

152. The film projector commonly used in extension

(a) 70 mm (b) 32 mm

(c) 16 mm (d) 8 mm

153. The poster used in demonstration should have a letter height of

(a) 2.0 inches (b) 1.5 inches

(c) 1.0 inches (d) 3.5 inches

154. The national press day is observed on

(a) 16th November (b) 5th June

(c) 06th May (d) 18th June

155. Media Asia is an important international Journal Published from

(a) USA (b) China

(c) Singapore (d) India

156. The name of the journal published from IIMC is

(a) Interaction (b) Vidura

(c) Communicator (d) Journal of Rural Development

157. The weekly agricultural column of "The Hindu" appears on

(a) Thursday (b) Sunday

(c) Monday (d) Friday

158. Rural India is published from

(a) Lucknow (b) Hyderabad

(c) Meerut (d) Pune

159. Agriculture and livestock is published by

(a) NIRD (b) ICAR

(c) MANAGE (d) UPCAR

160. The innovation decision process can be shortly called as

(a) MANAGE (b) KPDIC

(c) NAIDTEA (d) ATIC

161. Feedback is inferential in case of communication made through

(a) Block Level Worker (b) Group discussion

(c) Method Demonstration (d) Radio

162. Leagon's model of communication contains how many elements?

(a) 6 (b) 4

(c) 2 (d) 7

163. The book "Diffusion of innovations" is authored by

(a) Neil Roling (b) A.S.Sandhu

(c) Michael P. Collinson (d) E.M. Rogers

164. Aristotle model of communication contains how many ingredients?

(a) Two ingredients (b) Three ingredients

(c) Seven ingredients (d) Six ingredients

165. S and P waves are associated with

(a) Solar energy (b) Wind energy

(c) Tidal energy (d) Earthquakes

166. Black and white colours are

(a) Chromatic colours (b) Achromatic colours

(c) Diligent colours (d) Both (a) and (c)

167. " The Process of Mass Communication" was authored by

(a) Van den Ban (b) D Berlo

(c) Wilber Schramm (d) N. Roling

168. The first basic persuasive communication model

(a) Shannon and Weaver (b) Lazwell

(c) Aristotle (d) Berlo

169. Mathematical theory of communication was put forwarded by

(a) Shannon and Weaver *(b)* Lazwell
(c) Aristotle *(d)* Berlo

170. Non verbal communication based on touch is called

(a) Kinesis *(b)* Chronimics
(c) Haptics *(d)* Artifacts

171. The percentage of communication that takes place through body language

(a) 15 *(b)* 25
(c) 35 *(d)* 40

172. LCD expands to

(a) Laser Crystal Digital *(b)* Laser Crystal Display
(c) Laser Crystal Digital *(d)* Liquid Crystal Display

173. The Journal published from NIRD Hyderabad is

(a) Journal of Extension Education *(b)* Training and Development
(c) Journal of Rural Development *(d)* Journal of Extension

174. "Social action" is published from

(a) Institute of Social Sciences, New Delhi
(b) International Social Research Centre, New York
(c) Tata Institute of Social Research, Mumbai
(d) Indian Institute of Social Research, Kolkata

175. "Prasarika " is published from

(a) Maharashtra Society of Extension Education
(b) Indian Society of Extension Education
(c) Gujarat Society of Extension Education
(d) Rajasthan Society of Extension Education

176. "Communicator" is published from

(a) Indian Society of Extension Education, New Delhi
(b) Indian Institute of Mass Communication, New Delhi
(c) GBPUA&T, Pantnagar
(d) National Research Centre for Women in Agriculture, Bhubaneshwar

177. An official communication either oral or written is called

(a) Formal communication *(b)* Informal communication
(c) Both *(a)* and *(b)* *(d)* None of these

178. Who is the father of method demonstration ?

(*a*) Mr. Albert Mayer
(*b*) Dr. Seeman A Knapp
(*c*) Dr. Dadgil
(*d*) M. Henry Wolff

179. Audio aids are

(*a*) Radio
(*b*) Tape recorder
(*c*) Both (*a*) and (*b*)
(*d*) None of these

180. The chief emphasis in communication on

(*a*) Target channel
(*b*) Target communicator
(*c*) Target receiver
(*d*) Target audience

181. In which communication system of agencies works;

(*a*) Knowledge, Consuming and System
(*b*) Knowledge, Disseminating and System
(*c*) Knowledge, Generating and System
(*d*) None of these

182. Communication breakdown is a common feature of

(*a*) Straight transaction
(*b*) Vertical transaction
(*c*) Cross transaction
(*d*) None of these

183. The term intercommunication means

(*a*) Communication between two persons
(*b*) Communication between various groups
(*c*) Communication on any issue
(*d*) All of these

184. Flag method is classified under

(*a*) Group method
(*b*) Mass method
(*c*) Spoken method
(*d*) Written method

185. The term human communication means

(*a*) Exchange of ideas
(*b*) Exchange of service
(*c*) Exchange of liabilities
(*d*) None of these

186. Agricultural clinic is classified under

(*a*) Mass contact method
(*b*) Group contact method
(*c*) Individual contact method
(*d*) Both (*a*) and (*b*)

187. Which of the following is not a method of group communication ?

(a) Field Visits
(b) Conference
(c) Home Visits
(d) Nation Demonstration

188. In which communication system of mass media works

(a) Knowledge, Disseminating and System
(b) Knowledge, Consuming and System
(c) Knowledge, Generating and System
(d) None of these

189. The important points of result demonstration is/are

(a) Use audio-visual aid to support the result demonstration
(b) Never try to discover new truths
(c) Do not repeat demonstration repeatedly
(d) All of these

190. Educational television is classified under

(a) Screen media
(b) Printed media
(c) Broadcast media
(d) None of these

191. The authority of PACS is known as

(a) Secretary
(b) General Body
(c) Provident
(d) Executive Body

192. Under limited resource of a manpower time and money, we should select

(a) Man contact methods
(b) Individual contact methods
(c) Group contact methods
(d) All of these

193. The essence of communication is :

(a) Imparting knowledge
(b) Transmitting information
(c) Sharing information
(d) Sharing understanding

194. On which level, communication is fact and problem oriented

(a) Conventional level
(b) Participative level
(c) Explorative level
(d) None of these

195. The controlling of the flow of information through a communication channel

(a) Gate Security
(b) Gate guard
(c) Gate locking
(d) Gate keeping

196. A system of decision centers interconnected by patterned flows of information

(*a*) Communication network (*b*) Communication level

(*c*) Communication channel (*d*) Communication techniques

197. Among the following which is not a decentralized communication network?

(*a*) Wheel network (*b*) Star network

(*c*) Slash network (*d*) Circle network

198. Two linked individual in a network having personal communication networks that overlap

(*a*) Communication acceptance (*b*) Communication preference

(*c*) Communication proximity (*d*) None of these

199. The objectives of information and communication technology is/are

(*a*) To study the impact of ICT efforts on agriculture, health, women, empowerment and Panchayat Raj

(*b*) To print out the impediments for the effective implementation of ICT based project

(*c*) To elucidate the role of some of the major ICT initiatives in rural development

(*d*) All of these

200. The set of activities that facilitates by electronic means the capturing, storage, processing, transmission and display of intermation is

(*a*) Telecommunication technology

(*b*) Agricultural extension

(*c*) Communication

(*d*) Information and communication technology

FILL IN THE BLANKS

1. Telephone lady was introduced in ______________
2. International Telecommunication Union is headquartered at______
3. The full form of CeRA is ____________
4. AQUA Online expert Question and Answer based community forum, developed by____________
5. NABG-an ICAR initiative is associated with ____________
6. Google was founded by _________

7. WWW invented by ___________
8. The full form of http is ___________
9. The term computer is derived from the world___________
10. The Agricultural Libraries Network Coordinated by __________
11. Virtual Extension and Research Communication Network (VERCON) established by __________
12. Cybernetics is the science of __________
13. AKIS expands to __________
14. The toll free number on Kisan Call Centre is ___________
15. Kisan Call Centre started in the year___________
16. "Rice Doctor" an interactive pest diagnostic system is develop by ______
17. Krishi Darshan programme in India is telecasted from ______
18. Sponsoring agency of Agricultural Gateway of India is_______
19. Father of method Demonstration is ________
20. Round table discussion is called________

MARK TRUE OR FALSE

1. The e-choupal concept in extension delivery is promoted by ITC.
2. ATIC under NATP function under the control of ICAR.
3. Leaflets contain unlimited number of pages.
4. Pamphlets contain 12-24 number of pages.
6. Communication is the characteristic element in the Leagons Model.
7. An official communication either oral or written is known as formal communication.
8. The pioneering institution to draw upon folk music is Star Plus.
9. Weekly market plays an important role in the life of the tribal people.
10. Folk arts laying a meaningful role in rural areas in educating the urban people.
11. Audio aids are radio and tape recorder.
12. The mobile based e-agriculture initiative by TCS is M-Krishi.
13. BCC stands for blind carbon copy.
14. Linear model of communication model given by Aristotle.
15. Theory of communication was introduced by Lewin.
16. Uttar Pradesh state published large number of news paper.

17. Vividh Bharti launched in 1957
18. Communication I.Q. test was given by George Mead
19. Primary colours used in extension are Red, Blue, Yellow.
20. 'Communication' is published from Indian Institute of Mass Communication, New Delhi.

ANSWERS

Multiple Choice Questions

1.(b)	2.(c)	3.(c)	4.(a)	5.(b)	6.(a)	7.(a)
8.(c)	9.(b)	10.(a)	11.(a)	12.(d)	13.(c)	14.(d)
15.(a)	16.(d)	17.(c)	18.(d)	19.(d)	20.(d)	21.(a)
22.(a)	23.(c)	24.(c)	25.(a)	26.(c)	27.(b)	28.(a)
29.(b)	30.(d)	31.(a)	32.(b)	33.(c)	34.(a)	35.(d)
36.(a)	37.(c)	38.(a)	39.(c)	40.(a)	41.(d)	42.(a)
43.(c)	44.(b)	45.(a)	46.(a)	47.(a)	48.(c)	49.(b)
50.(a)	51.(d)	52.(a)	53.(d)	54.(b)	55.(a)	56.(b)
57.(c)	58.(d)	59.(c)	60.(a)	61.(c)	62.(c)	63.(b)
64.(a)	65.(d)	66.(c)	67.(d)	68.(a)	69.(c)	70.(d)
71.(c)	72.(d)	73.(d)	74.(c)	75.(c)	76.(c)	77.(c)
78.(a)	79.(b)	80.(b)	81.(b)	82.(c)	83.(c)	84.(b)
85.(b)	86.(c)	87.(b)	88.(c)	89.(a)	90.(c)	91.(c)
92.(c)	93.(b)	94.(a)	95.(c)	96.(b)	97.(b)	98.(d)
99.(b)	100.(a)	101.(b)	102.(c)	103.(c)	104.(a)	105.(d)
106.(a)	107.(b)	108.(b)	109.(a)	110.(a)	111.(d)	112.(c)
113.(c)	114.(c)	115.(c)	116.(d)	117.(c)	118.(a)	119.(a)
120.(c)	121.(c)	122.(a)	123.(d)	124.(a)	125.(b)	126.(d)
127.(d)	128.(a)	129.(c)	130.(d)	131.(a)	132.(b)	133.(a)
134.(a)	135.(d)	136.(b)	137.(b)	138.(c)	139.(c)	140.(c)
141.(d)	142.(d)	143.(d)	144.(a)	145.(d)	146.(a)	147.(a)
148.(a)	149.(a)	150.(c)	151.(c)	152.(c)	153.(b)	154.(a)
155.(c)	156.(c)	157.(a)	158.(d)	159.(b)	160.(b)	161.(d)
162.(a)	163.(d)	164.(b)	165.(d)	166.(b)	167.(b)	168.(c)
169.(a)	170.(c)	171.(d)	172.(d)	173.(c)	174.(a)	175.(d)
176.(b)	177.(a)	178.(b)	179.(c)	180.(d)	181.(c)	182.(c)
183.(d)	184.(c)	185(a)	186.(c)	187.(b)	188.(a)	189.(d)
190.(c)	191.(a)	192.(a)	193.(d)	194.(c)	195.(d)	196.(a)
197.(b)	198.(c)	199.(d)	200.(d)			

Fill in the Blanks

1. Bangladesh
2. Geneva
3. The Constorium for e-resources in Agriculture
4. Media Lab Asia and IIT Mumbai
5. Bioinformatics
6. Larry Page and Sergey Brin
7. Tim Bernners-Lee
8. Hyper Text Transfer Protocol
9. Compute
10. FAO
11. FAO
12. Communication
13. Agricultural Knowledge and Information System
14. 1551
15. 1972
16. IRRI
17. New Delhi
18. NAARM, Hyderabad
19. Dr Seeman A Knapp
20. Panel

Mark True or False

1. True
2. True
3. False
4. True
5. False
6. True
7. True
8. False
9. True
10. False
11. True
12. True
13. True
14. True
15. False
16. True
17. True
18. False
19. True
20. True

CHAPTER 3

PROGRAMME, PLANNING AND RURAL DEVELOPMENT IN AGRICULTURE

Multiple Choice Questions

1. **Indian Society of Extension Education was established on**
   (a) 17th January 1967 (b) 22nd June 1964
   (c) 18th November 1966 (d) 11th September 1968
2. **The first extension education institute in India is established in**
   (a) Hyderabad (b) Anand
   (c) Nilokari (d) Pantnagar
3. **Agricultural Scientist Recruitment Board (ASRB) was initiated in the year**
   (a) 1975 (b) 1970
   (c) 1973 (d) 1978
4. **Which KVK zone has maximum number of KVK in India ?**
   (a) Zone II (b) Zone IV
   (c) Zone VII (d) Zone V
5. **Which state has maximum KVK ?**
   (a) Bihar (b) UP
   (c) Rajasthan (d) Tamil Nadu
6. **Department of Rural Development was established by Government of India in**
   (a) 1921 (b) 1931
   (c) 1947 (d) 1944

7. Sewagram project was started by Mahatma Gandhi in the state of

(a) Gujarat (b) Madhya Pradesh

(c) Rajasthan (d) Bihar

8. Sarvodya Scheme was started in 1948 by

(a) Acharya Vinoba Bhave (b) Mahatma Gandhi

(c) O.P. Dhama (d) Ravindra Nath Tagore

9. The number of villages in which Etawah pilot was initially started in 1948 was

(a) 48 (b) 64

(c) 88 (d) 17

10. In Etawah pilot project, the headquarters were established at

(a) Eeta (b) Gurgaon

(c) Ameti (d) Mahewa

11. Nilokheri project was also called

(a) Mazdoor Manzil (b) Kishan Manzil

(c) Mazdoor Madel (d) None of these

12. Community Development project was started as a result of

(a) Educational Commission Report (b) Royal Commission Report

(c) GMF Enquiry Committee Report (d) None of these

13. Community Development Programme was transferred to the state sector on

(a) 16 February, 1967 (b) 18 November, 1970

(c) 01 April, 1969 (d) 15 May, 1975

14. Which of the following programme was developed on the basis of "10 – point programme"?

(a) CDP (b) NES

(c) IAAP (d) IADP

15. The National Dairy Development Board (NDDB) was registered as an autonomous government society in

(a) 1963 (b) 1965

(c) 1971 (d) 1980

16. Who is called the "Father of White Revolution in India"?

(a) Verghese Kurien (b) B.P. Pal

(c) M.S. Swaminathan (d) David Gibbo

17. The agency responsible for planning, implementation coordination, supervision and monitoring of IRDP at the district level is

(a) Ministry of Rural Development (b) DRDA

(c) NABARD (d) None of these

18. A technical committee was constituted in 1993 to review Drought Prone Area Programme (DPAP) and Desert Development Programme (DDP) under the chairmanship of

(a) B.P. Pal (b) R.S. Paroda

(c) M.S. Swaminathan (d) C.H. Hanumantha Rao

19. "Operation Flood" is related to

(a) Linking of rivers (b) Controlling flood in N-E states

(c) Dairy development (d) Indus river treaty

20. In 1995, the international Cooperative Alliance (ICA) Congress was held at

(a) Manchester (b) Geneva

(c) New York (d) New Delhi

21. The financial institution set up a result of recommendation of "Committee to Review Arrangements for Institutional Credit for Agriculture and Rural Development (CRAFICAR)

(a) RRBs (b) NABARD

(c) Cooperative Banks (d) Bank Nationalization

22. The University Education Commission of 1949 which recommended the establishment of "Rural Universities" was headed by

(a) Dr. Mohan Sinha Mehta (b) Dr. S. Radhakrishnan

(c) Dr. M.S. Swaminathan (d) Dr. B.P. Pal

23. National Institute of Agriculture Extension Management (MANAGE) was set in

(a) 1984 (b) 1987

(c) 1986 (d) 1982

24. A person who organizes and runs a business enterprise is responsible and makes profit is more perfectly called

(a) An entrepreneur (b) A businessman

(c) A merchant (d) Shop Keeper

25. The term evaluation is derived from

(*a*) Greek (*b*) English

(*c*) German (*d*) Latin

26. CAPART stands for

(*a*) Council for Action and People Advancement in Rural Technology

(*b*) Council in Acting the Participatory Advancement of Rural Technology

(*c*) Council for Active involvement of People in Ruralite Technology

(*d*) Council for Advancement of People Action and Rural Technology

27. CAPART is an autonomous wing of

(*a*) Ministry of Agriculture and Farmers Welfare

(*b*) Ministry of Human Resource Development

(*c*) Ministry of Rural Development

(*d*) Ministry of Finance

28. The headquarters of International Service for National Agricultural Research (ISNAR) is located in

(*a*) Mexico (*b*) USA

(*c*) China (*d*) The Netherlands

29. NAARM is situated at

(*a*) New Delhi (*b*) Hyderabad

(*c*) Bikaner (*d*) Lucknow

30. The first newspaper started in India was

(*a*) Rural India (*b*) Rural –Urban

(*c*) Bengal outlook (*d*) Bengal Gazette

31. "Bengal Gazette" was a

(*a*) Daily (*b*) Weekly

(*c*) Monthly (*d*) Bi-monthly

32. The National Service Scheme (NSS) was launched in the country in the year

(*a*) 1971 (*b*) 1969

(*c*) 1960 (*d*) 1951

33. Young Farmers Association (YFA) was founded under the guidance of erstwhile Union Minister of Agriculture

(*a*) P.S. Deshmukh (*b*) C.P. Thakur

(*c*) G.K. Puranyak (*d*) None of these

34. The Headquarters of International Farm Youth exchange programme is at

(a) New York (b) Washington

(c) China (d) Addis Ababa

35. "Krishi –Dharshan" programme was launched in India on

(a) 26th January 1961 (b) 26th January 1967

(c) 15th August 1964 (d) 15th August 1977

36. Television in India started on

(a) August 15th 1956 (b) August 15th 1958

(c) January 26th 1957 (d) None of these

37. The concept of Participatory Development Approach (PDA) basically originated from

(a) Political movements (b) Economic Movement

(c) **Social movements** (d) Geographical dislocation

38. The book "Participatory Rural Appraisal–Methodology and Applications " is written by

(a) M.S. Swaminathan (b) Neela Mukherjee

(c) Baldeo Singh (d) R.N. Tagore

39. National Research Center for Women in Agriculture (NRCWA) has been set up by

(a) Ministry of Agriculture and Farmers Welfare

(b) Ministry of Human Resource Development

(c) Indian Council of Agricultural Research

(d) Indian Council of Sociological Research

40. The basic orientation of Programme Evaluation and Reviewing Technique (PERT) is

(a) Opportunistic (b) Deterministic

(c) Probabilistic (d) None of these

41. The networking technique applied to the projects that employ a fairly and risk free technology

(a) PERT (b) CPM

(c) PROMPT (d) POP

42 Division of Extension was set up in India under the expert guidance of

(a) Albert Mayer (b) J.P. Leagons

(c) John Dewey (d) Paul Preston

43. The first institute in India to start teaching of extension education at undergraduate level was

(*a*) College of Agriculture, New Delhi

(*b*) College of Agriculture, Madras University

(*c*) College of Agriculture, GB Pant University

(*d*) College of Agriculture, Calcutta University

44. The first post graduate programme in extension education was started in India in the year

(*a*) 1945 (*b*) 1955

(*c*) 1953 (*d*) 1960

45. ATMA is an/a

(*a*) Registered society (*b*) Unregistered society

(*c*) Recognized society (*d*) None of these

46. The full form ATMA is

(*a*) Agricultural Technical Management Association

(*b*) Agricultural Technology Management Agency

(*c*) Associated Techniques for Management of Agriculture

(*d*) None of these

47. SREP stands for

(*a*) Strategic Rural Employment Plan

(*b*) Source Receiver Evaluation Plan

(*c*) Strategic Research Extension Plan

(*d*) Strategic Research Evaluation Programme

48. The concept of ATMA has been introduced as a/an

(*a*) Conservative organization (*b*) Descriptive organization

(*c*) Semi-autonomous organization (*d*) Autonomous organization

49. ATMA Governing Board is a/an

(*a*) Statutory body (*b*) Advisory body

(*c*) Autonomous body (*d*) None of these

50. The full form of ATIC is

(*a*) Agricultural Technology Initiation Committee

(*b*) Agriculture Technology Information Committee

(*c*) Agricultural Technical Initiation Center

(*d*) Agricultural Technology Information Center

51. ATICs were set up in India in the year

(a) 2000 (b) 1998

(c) 1996 (d) 1981

52. ATICs were set up by

(a) Ministry of Agriculture, GOI

(b) Ministry of Rural Development, GOI

(c) State Department of Agriculture

(d) Indian Council of Agricultural Research

53. ATICs are under the administrative control of

(a) Deputy Director Extension, ICAR

(b) Director General, ICAR

(c) Minister of State for Agriculture

(d) Deputy Director Education, ICAR

54. IVLP stands for

(a) Institutional Village Linkage Programme

(b) Institutional Village Linked Programme

(c) Institutional Village Likelihood Project

(d) None of these

55. The type of services provided by agri-clinics and agri-business centers to the farmer is

(a) Voluntary (b) Free

(c) Paid (d) None

56. Doordarshan started using INSAT for national wide transmission of channels in

(a) 1984 (b) 1982

(c) 1985 (d) 1980

57. The broadcasting services in India were known to be All India Radio (AIR) in the year

(a) 1934 (b) 1936

(c) 1940 (d) 1937

58. The first telecast on Television in India was started on

(a) 15th November 1958 (b) 15th October 1947

(c) 15th September 1959 (d) 15th August 1954

59. Television programme was first telecast from

(*a*) Mumbai (*b*) Chennai

(*c*) New Delhi (*d*) Kolkata

60. Which of the following programme was the first attempt in the world at using the sophisticated technology of satellite transmission for social education?

(*a*) SITE (*b*) SCRES

(*c*) STEP (*d*) ATIC

61. DD-Gyandarshan, an exclusive educational channel was started in India on

(*a*) 26th January 1995 (*b*) 26th January 2000

(*c*) 15th August 1988 (*d*) 15th August 1976

62. From among which of the following states large number of newspapers are published

(*a*) Uttar Pradesh (*b*) Madhya Pradesh

(*c*) Rajasthan (*d*) Maharashtra

63. The National Press Day is observed on

(*a*) 16 November (*b*) 5 June

(*c*) 15 August (*d*) 31 May

64. The ICRISAT stands for

(*a*) Indian Crop Research Institute for Semi-Arid Tropics

(*b*) Indian Centre for Research on Input Supply of Agriculture Technology

(*c*) International Council for Research Institute for Semi-Arid Tropics

(*d*) International Crop Research Institute for Semi-Arid Tropics

65. The headquarters of "Food and Agricultural Organization" are at

(*a*) New York (*b*) New Delhi

(*c*) Rome (*d*) China

66. The full form of NAARM is

(*a*) National Association of Agricultural Research Management

(*b*) National Academy of Agricultural Research Management

(*c*) National Academy of Agricultural Resource Management

(*d*) None of these

67. The first KVK was established in 1974 at

(a) Pondichery (b) Nilokheri
(c) Ludhiana (d) Hyderabad

68. Purpose of extension evaluation

(a) To identify the weak points (b) To identify the strong points
(c) To identify the gaps and errors (d) All of these

69. Lab to land programme was started by

(a) ICAR (b) Govt. of Tamil Nadu
(c) Smt. Indira Gandhi (d) Govt. of Rajasthan

70. People's participation in an extension programme is significant when

(a) Local leaders participate
(b) Literate section of villagers participate
(c) Gram Panchayat members participate
(d) Majority of villagers participate

71. Success in rural development project generally depends upon

(a) Regular training of workers (b) Regular contact of workers
(c) Both (a) and (b) (d) Participation of beneficiaries

72. National Academy of Agriculture Research Management is situated at

(a) Hyderabad (b) New Delhi
(c) Bangalore (d) Pantnagar

73. The last state in the country which ultimately decided to introduce VAT for replacing trade tax is ______

(a) Uttar Pradesh (b) Tripura
(c) Meghalaya (d) None of these

74. The equity of Reserve Bank of India in national housing bank is

(a) 39 per cent (b) 80 per cent
(c) 71 per cent (d) 100 per cent

75. World habitat day is observed on

(a) 15 September (b) 30 September
(c) 1 October (d) 5 June

76. National Association of Software and Services Companies (NASSCOM) founded in the year

(a) March 1975 (b) March 1978

(c) March 1985 (d) March 1988

77. KVKs are formed to offer

(a) In-service training (b) On the job training

(c) Vocational training (d) All of these

78. National demonstration is an example for

(a) Future line demonstration (b) First line demonstration

(c) Front line demonstration (d) None of these

79. The Funding of KVK is done by

(a) SAU (b) ICAR

(c) Central government (d) State government

80. Farm science is a synonym of

(a) KVK (b) FDG

(c) FTC (d) Teleclub

81. Krishi Gyan Kendra function with the financial assistance of

(a) State Government (b) Central Government

(c) ICAR (d) MANAGE

82. NARS stands for

(a) National Applied Research System

(b) National Agricultural Resource System

(c) National Agricultural Revenue System

(d) National Agricultural Research System

83. TARP stands for

(a) Technology Awareness and Refinement Project

(b) Technology Adoption and Refinement Project

(c) Technology Assessment and Reinvention Project

(d) Technology Assessment and Refinement Project

84. Which is the largest self employment programme for rural poor ?

(a) SJSRY (b) SGSY

(c) NREP (d) All of these

85. The reservation for women as beneficiary under SGSY is

(*a*) 33.33 percent (*b*) 40 percent
(*c*) 50 percent (*d*) 10 percent

86. The reservation for SC/ST as beneficiary under SGSY is

(*a*) 33.33 per cent (*b*) 10 per cent
(*c*) 50 per cent (*d*) 63.33 per cent

87. Swarna Jayanti Shahari Rozgar Yojna (SGSRY) was launched in India from

(*a*) 2nd October 1995 (*b*) 2nd October 2001
(*c*) 14th November 2001 (*d*) 1st December 1997

88. Mid Day Meal Scheme covers children studying at

(*a*) Collegiate level (*b*) High school level
(*c*) Secondary level (*d*) Primary level

89. Day of elderly was observed on

(*a*) May 31 (*b*) June 5
(*c*) October 1 (*d*) December 10

90. Jawahar Rozgar Yojana was started by

(*a*) Jawaharlal Nehru (*b*) Rajiv Gandhi
(*c*) Dr. Manmohan Singh (*d*) Indira Gandhi

91. In which year National adult education programme was launched ?

(*a*) 1988 (*b*) 1977
(*c*) 1987 (*d*) 1972

92. Operation black board aims to improve

(*a*) Professional education (*b*) Informal education
(*c*) Adult education (*d*) Primary education

93. Consequent upon the policy to appoint an agriculture scientist as the chief executive of Indian council of agricultural research (ICAR), in the year 1965 the assignment as the first director general of ICAR was given to

(*a*) Dr. M.S. Swaminathan (*b*) Dr. B.P. Pal
(*c*) Sir Jogendra Singh (*d*) Dr. K.L. Chadha

94. Who was the first president of imperial council of agricultural research, the present day Indian council of agricultural research?

(*a*) Sir Mohammad Habibullah (*b*) Sir G.B. Nayak
(*c*) Sir Kamal Pandey (*d*) None of these

95. In which year PIRCOM was initiated ?

(*a*) 1952 (*b*) 1956

(*c*) 1940 (*d*) 1975

96. PIRCOM is stands for

(*a*) Project and input research center on management of input services

(*b*) Planning and implementation of regionally common multipurpose input system

(*c*) Project for intensification of regional research on cotton, oilseeds and millets

(*d*) None of these

97. MANAGE functions under

(*a*) Ministry of Agriculture and Farmers Welfare

(*b*) Ministry of Tribal Affairs

(*c*) Ministry of Human Resource Development

(*d*) Ministry of Rural Development

98. NIRD is under

(*a*) Ministry of Agriculture and Farmers Welfare

(*b*) Ministry of Tribal Affairs

(*c*) Ministry of Human Resource Development

(*d*) Ministry of Rural Development

99. Who was set up IARI in Bihar in 1905 ?

(*a*) Lord Curzon (*b*) Lord Doulhossie

(*c*) Lord Minto (*d*) Lord Hardinge

100. India is a member of GATT since

(*a*) 1947 (*b*) 1960

(*c*) 1950 (*d*) 1971

101. The TRIIMS is stands for

(*a*) Trade Relation interest Means

(*b*) Trade Related Investment Measures

(*c*) Trade and Research in Management Systems

(*d*) None of these

102. The full form of TRIPS is

(*a*) Trade Related Information Policy System

(*b*) Trade Related Intellectual Property Rights

(*c*) Trade Related to information and Processing

(*d*) None of these

103. A good extension programme should be

(*a*) Flexible (*b*) Rigid

(*c*) Both (*a*) and (*b*) (*d*) None

104. Basic operational unit for rural development in India is

(*a*) Village (*b*) District

(*c*) State (*d*) Block

105. Crop demonstration were first laid in 1903 by

(*a*) Seaman A. Knapp (*b*) M.S. Mehta

(*c*) Daniel Benor (*d*) A.T. Mosher

106. The concept of Farmer First was extensively promoted by

(*a*) Neils Rolling (*b*) Robert Chambers

(*c*) R.S. Paroda (*d*) B.P. Pal

107. The collection of data in RRA is done by

(*a*) Local extension agents (*b*) Progressive farmers

(*d*) Research assistants (*d*) A Multidisciplinary team

108. The programme TRYSEM was merged under the cover of single umbrella known as

(*a*) Swarna Jayanti Rozgar Yojana

(*b*) Jawahar Rozgar Yojana

(*c*) Integrated Rural Development Programme

(*d*) Green Revolution

109. Who among the following is the chairman of NITI Aayog?

(*a*) Prime Minister (*b*) Finance Minister

(*c*) Secretary to NITI Aayog (*d*) HRD Minister

110. An outline of activities so arranged so as to enable efficient execution of the programme

(*a*) Calendar of work (*b*) Plan of work

(*c*) Plan (*d*) Project

111. A predetermined course of action is known as

(a) Objective (b) Project

(c) Plan (d) Management

112. Agricultural Scientist Recruitment Board (ASRB) was set up in the year

(a) 1971 (b) 1976

(c) 1970 (d) 1973

113. The Key Village Scheme comprising first systematic attempt to improve quality and productivity of cattle and buffaloes in India

(a) July, 1952 (b) June, 1958

(c) July, 1957 (d) August, 1952

114. Farmer First (FF) Model was developed in late eighties by the institute of Development Studies, University of Sussex, in

(a) UK (b) USA

(c) China (d) India

115. The "Kisan Credit Card Scheme" was introduced in the year

(a) 1995-96 (b) 1990-91

(c) 1998-99 (d) 2000-2002

116. Kisan Credit Scheme was launched in 1998-99 by

(a) PNB (b) SBI

(c) IOB (d) BOI

117. The term "Green Revolution" was coined by

(a) Dr. Yuvan Long Ping (b) Dr. M.S. Swaminathan

(c) William Gadd (d) Dr. B.P. Pal

118. Under the modified T and V system "Kisan Seva Kendra"were set up at

(a) Sub-divisional Agriculture Office (b) District Agriculture Office

(c) Headquarters of VEW (d) Headquarters of AEO

119. Lab to Land Programme was initially started for the period of

(a) 1 year (b) 6 years

(c) 4 years (d) 7 years

120. Who was started Jawahar Rozgar Yojana ?

(a) Indira Gandhi (b) Rajeev Gandhi

(c) A.B.Vajpayee (d) Chandra Sekhar Singh

121. In 1988, the Rural System Research (RSR) idea was motivated by

(a) Ralph W. Cummings (b) Guy B. Baird

(c) M.S. Swaminathan (d) Normal E. Borlaoug

122. The 4-H national center was set up in Washington through a Ford Foundation grant in the year

(a) 1586 (b) 1950

(c) 1956 (d) 1959

123. Subsequent to the nationalization of banks in 1969,six more commercial banks were nationalized in

(a) 19th July, 1972 (b) 15th April, 1980

(c) 20th October, 1970 (d) 18th June, 1975

124. Balwant Rai Mehta Team stimulated on active consideration of ______ through democratic bodies

(a) Centralization (b) Decentralization

(c) Mobilization (d) None of these

125. A leaflet is a single sheet of paper folded usually to make ______ page piece of printed paper

(a) Five (b) Three

(c) Four (d) Eight

126. The first Agricultural University in India came into existence in 1960 at

(a) Ludhina, Punjab (b) Pantnagar, Uttarakhand

(c) Rahuri, Maharashtra (d) Kanpur, Uttar Pradesh

127. The programmes, which have been merged under IRDP, were

(a) SFDA, MFAL, DPAP and CADP

(b) MFAL, SFDA, IAAP AND IADP

(c) SFDA, MFAL AND Lab TO Land

(d) None of the above

128. For the implementation of CDP, the Technical Cooperation Programme Agreement was made between Government of India and Government of

(a) **USA** (b) UK

(c) Israel (d) Mexico

129. What was" India's food crisis and steps to meet it"?

(a) A book on food situation in India

(b) Report of Ford Foundation Team

(c) Report related to poverty in India

(d) A formula to solve food disease in India

130. The central idea behind IADP was increased agriculture production should lead to ______ which shall bring welfare to the society

(a) Social growth
(b) Economic growth
(c) Personal growth
(d) None of these

131. "The Journal of Cooperative Extension Service" is published from

(a) USA
(b) China
(c) India
(d) Mexico

132. In which language the newspapers have the highest circulation?

(a) Hindi
(b) English
(c) Bengali
(d) Gujarati

133. The constitutional amendment made to ensure free and compulsory education to all children

(a) 92nd Amendment
(b) 86th Amendment
(c) 74th Amendment
(d) 78rd Amendment

134. Free and compulsory education to all children in the age group of 6- 14 years is envisaged in

(a) Kasturba Gandhi Balika Vidyalaya

(b) Sarva Shiksha Abhiyan

(c) National Literacy Mission

(d) national Adult Education Programme

135. Sarva Shiksha Abhiyan was started in the year

(a) 2002
(b) 2005
(c) 2000
(d) 1994

136. Kasturba Gandhi Balika Vidyalaya is meant for promoting education of

(a) Agricultural labourers
(b) Rural youth
(c) Senior citizen
(d) Girl children

137. 'Files to field programme has been introduced in

(a) New Delhi (b) Ahmedabad

(c) Lucknow (d) Ahmednagar

138. Who is the head of Zila Parishad?

(a) Collector (b) Chief executive officer

(c) Agriculture development officer (d) None of these

139. Which type of training center is Krishi Vigyan Kendra?

(a) Trainers' training center (b) Extension training center

(c) Rural leadership training center (d) Vocational training center

140. In which district of Rajasthan, IADP was initially launched?

(a) Pali (b) Bharatpur

(c) Kota (d) Bikaner

141. The Directorate of extension at the national level is located in the Ministry of

(a) Agriculture and Farmer's Welfare (b) Communication

(c) Human Resource Development (d) Rural Development

142. The extension work must be based on the needs and interests of

(a) People (b) Government

(c) Extension worker (d) None of these

143. "Operation flood" programme relates to

(a) Conserving soil (b) Boosting fish production

(c) Funds for flood victims (d) Boosting milk supply

144. Department of Agricultural Research and Education (DARE) was set up under the Ministry of Food and Agriculture in the year

(a) 1971 (b) 1973

(c) 1965 (d) 1970

145. National Institute of Science Communication and Information Resources (NISCAIR) is located at

(a) Hyderabad (b) Bikaner

(c) New Delhi (d) Lucknow

146. Which article of agreement on agriculture (AoA) of WTO specifically provides for the process for agricultural sector trade reforms

(a) Article 20 (b) Article 20

(c) Article 25 (d) Article 26

147. National Rural Employment Programme (NREP) was replaced to food and work programme in

(*a*) 1982 (*b*) 1971

(*c*) 1980 (*d*) 1970

148. The Institution of Village Panchayat exist in all the union territories expect

(*a*) New Delhi (*b*) Chandigarh

(*c*) Lakshadweep and Mizoram (*d*) None of these

149. National Development Council (NDC)is headed by

(*a*) Prime Minister (*b*) Home Minister

(*c*) HRD Minister (*d*) Agriculture and Farmers Welfare

150. Nationalization of commercial banks happened in the year

(*a*) 1971 (*b*) 1969

(*c*) 1981 (*d*) 1970

151. "Each for all and all for each" forms the basic principle of

(*a*) NABARD (*b*) ATMA

(*c*) Planning Commission (*d*) Cooperative Banks

152. National Agricultural Cooperative Marketing Federation on India Ltd. has its headquarters at

(*a*) Bikaner (*b*) Hyderabad

(*c*) Lucknow (*d*) New Delhi

153.. The first cooperative society was started in the year

(*a*) 1937 (*b*) 1910

(*c*) 1912 (*d*) 1904

154. The minimum members required to register a co-operative society

(*a*) 25 (*b*) 24

(*c*) 16 (*d*) 10

155. What is Gross Domestic Product (GDP) ?

(*a*) The total of goods and services produced by a nation over a given period, usually 1 year

(*b*) The value of all final goods and services produced within a nation in a given period

(*c*) Value if expenditure

(*d*) None of above

156. Boodhan Movement was started by

(a) Acharya Vinoba Bhave
(b) Mahatma Gandhi
(c) R.C.Reddy
(d) None of these

157. Central Social Welfare Board was set up in

(a) July 1958
(b) October 1953
(c) August 1960
(d) November 1957

158. President of Ford Foundation who came to India in 1951

(a) Paul Hoffman
(b) Priestly Comity
(c) Royal Paul Entrust
(d) None of these

159. National Extension Services was started on 2nd of October

(a) 1960
(b) 1953
(c) 1955
(d) 1957

160. The idea behind the National Extension Services (NES) was to cover the entire country with in the period of

(a) 3 Years
(b) 9 Years
(c) 10 Years
(d) 12 Years

161. Lab to land programme was started in the country as a part of ICAR ______ Jubilee celebrations

(a) Golden
(b) Silver
(c) Diamond
(d) Platinum

162. The Government of India established Department of Agriculture in

(a) 1901
(b) 1905
(c) 1871
(d) 1890

163. Imperial Agriculture Research Institute (Now Indian Agriculture Research Institute) was established in

(a) 1900
(b) 1907
(c) 1905
(d) 1904

164. Sriniketan project was started in the groups of how many villages of Calcutta

(a) 9
(b) 7
(c) 8
(d) 6

165. Sriniketan project was started in Calcutta by

(a) Mahatma Gandhi
(b) R.N. Tagore
(c) Vinobha Bhave
(d) P.V. Young

166. The word "Democracy " is derived from Greek roots *'Demos'* meaning ______ and *"Kartos"* meaning rule/authority.

(*a*) King (*b*) People

(*c*) Leader (*d*) None of these

167. The Democratic Decentralization was recommended by the team on Community Development Programmes headed by

(*a*) Khuswant Singh Mehta (*b*) Balwant Raj Mehta

(*c*) G.R.Mehat (*d*) None of these

168. In which year Lead bank scheme was introduced ?

(*a*) 1975 (*b*) 1970

(*c*) 1969 (*d*) 1965

169. Regional Rural Bank was set up in the year

(*a*) 1975 (*b*) 1971

(*c*) 1961 (*d*) 1935

170. Railway Recruitment Board (RRB) was set up under the recommendation of

(*a*) Nariman committee (*b*) Sen committee

(*c*) Narasimham committee (*d*) None of these

171. The first bank to launch Kisan Credit Card

(*a*) Punjab National Bank (*b*) Cooperation Bank

(*c*) State Bank of India (*d*) Bank of India

172. The recently launched weather insurance scheme of government of India

(*a*) Varsha Beema (*b*) Ganga Beema

(*c*) Barsat Beema (*d*) None of these

173. In which State of India the Panchayat Raj system was firstly started ?

(*a*) Rajasthan (*b*) Uttar Pradesh

(*c*) Haryana (*d*) Tamil Nadu

174. The Right to Information Act was passed in the year

(*a*) 2001 (*b*) 2004

(*c*) 2003 (*d*) 2005

175. World Habitat Day is celebrated on first Monday of

(*a*) May (*b*) November

(*c*) October (*d*) December

176. Mission 2007 refers to

(a) Medicinal plants
(b) Microirrigation
(c) Knowledge centers
(d) None of these

177. The recent National Commission on Agriculture was headed by

(a) Sharad Pawar
(b) Pramod Tandon
(c) Mangala Rai
(d) M.S. Swaminathan
(c) Ministry of Human Resource Development
(d) Ministry of Rural Development

178. Indian Council of Agricultural Research (ICAR) was set up in 1929 on the basis of the report of

(a) Nalagarh Committee
(b) Royal Commission
(c) Famine Commission of 1901
(d) None of these

179. Imperial Council of Agriculture Research (Now Indian Council of Agricultural Research) was established in

(a) 1910
(b) 1940
(c) 1929
(d) 1937

180. Who among the following is associated with YMCA

(a) Kuldeep Nair
(b) Spencer Hatch
(c) Edger Dale
(d) Gabriel Trade

181. Panchayat Raj was introduced on the basis of the recommendations of

(a) Grow More Food Enquiry Committee
(b) V.T. Krishnamachari Committee
(c) Royal Commission of Agriculture
(d) Balwant Raj Mehta Committee

182. During the first Five Year plan, the highest priority was accorded to

(a) Agriculture
(b) Industries
(c) Defense
(d) Rural Development

183. The Intensive Agriculture District Programme (IADP) was popularly known as

(a) Training programme
(b) Mass contact programme
(c) Package programme
(d) All of these

184. The international Institute of Rural Reconstruction is located at

(*a*) Hyderabad (*b*) Cuttack

(*c*) New Delhi (*d*) **Philippines**

185. The first institution to start M.Sc. teaching in extension education was located at

(*a*) Sabour (*b*) Bikaner

(*c*) New Delhi (*d*) Pantnagar

186. According to Keisey and Hearne, how many steps are there in extension programme planning?

(*a*) Eight (*b*) Seven

(*c*) Six (*d*) Twelve

187. Identification of needs of people is part of

(*a*) Programme of work (*b*) Step of demonstration

(*c*) Best quality of work (*d*) Teaching method

188. Who, what, when and where are part of

(*a*) Method of approach (*b*) Calendar of work

(*c*) Plan of Work (*d*) Evaluation

189. HYVP is related to

(*a*) Animal husbandry (*b*) Popularizing crops

(*c*) Milk production (*d*) Fish Production

190. The Ph.D. teaching in extension education was started in the year

(*a*) 1961 (*b*) 1975

(*c*) 1971 (*d*) 1970

191. Food grains production includes the production of

(*a*) Cereals only (*b*) Pulses only

(*c*) Cereals and Pulses (*d*) Cereals, Pulses and Oilseeds

192. The T and V system of extension was first introduced in

(*a*) Turkey (*b*) China

(*c*) Mexico (*d*) India

193. The participation of people in an extension programme is

(*a*) Compulsory (*b*) Voluntary

(*c*) Involuntary (*d*) Both (*a*) and (*c*)

194. Community development project was started in India in the year

(*a*) 1952 (*b*) 1960

(*c*) 1972 (*d*) 1981

195. ATMA is stands for

(*a*) Agriculture Technology Management Association

(*b*) Agriculture Technology Management Agency

(*c*) Agriculture Technology Mission Agency

(*d*) None of the above

196. Who was started T and V systems of extension ?

(*a*) D. Benor (*b*) Marx Weber

(*c*) A.H. Maslow (*d*) Mc Gregor

197. In which year high yielding variety programme was started?

(*a*) 70-81 (*b*) 1965-66

(*c*) 1956-60 (*d*) 1971-1975

198. "Kisan Bharti" periodical is published from

(*a*) Pant Nagar (*b*) Hyderabad

(*c*) New Delhi (*d*) Lucknow

199. DRDA is established at

(*a*) State level (*b*) Block level

(*c*) Village level (*d*) District level

200. Panchayat Raj report was submitted by Balwant Raj Mehta in the year

(*a*) 1981 (*b*) 1957

(*c*) 1951 (*d*) 1965

201. The impact of "green revolution" was largely noted in

(*a*) Rice only (*b*) Rice and wheat only

(*c*) Wheat (*d*) Pulses only

202. The department of agriculture was set up in India during the tenure of

(*a*) Lord Irwin (*b*) Lord Minto

(*c*) Lord Canning (*d*) Lord Mayo

203. Royal commission on agriculture was headed by

(*a*) Lord Minto (*b*) Lord Linlithgow

(*c*) Lord Defferon (*d*) Lord Hardinge

204. NATP is funded by

(*a*) ICAR (*b*) FAO

(*c*) SAU's (*d*) World Bank

205. NAIP stands for

(*a*) National Agricultural Invitation Project

(*b*) National Agricultural Innovation Programme

(*c*) National Agricultural Innovation Project

(*d*) None of these

206. The age group of the members of 4-H club is

(*a*) 15-25 (*b*) 5-10

(*c*) 15-20 (*d*) 10-20

207. Massagana 99, a successful transfer of technology programme of rice is implemented

(*a*) Germany (*b*) China

(*c*) India (*d*) Philippines

208. BRAC, one of the world's largest voluntary organization, have its base at

(*a*) China (*b*) Srilanka

(*c*) Bangladesh (*d*) Philippines

209. Panchayat Raj institutions are formed based on

(*a*) Ashok Mehta Committee (*b*) Hanumantha Rao Committee

(*c*) Balwanthray Mehta Committee (*d*) None of these

210. Panchayat Raj is a system of

(*a*) Local government (*b*) Local self-government

(*c*) Local administration (*d*) None of these

211. The term Panchayat Raj was christened by

(*a*) J.L. Nehru (*b*) M.S. Mehta

(*c*) M.K. Gandhi (*d*) None of these

212. In which year Panchayat Raj was introduced in India ?

(*a*) 1972 (*b*) 1958

(*c*) 1958 (*d*) 1955

213. NES was considered basically as a

(*a*) Agency (*b*) Method

(*c*) Project (*d*) Report

214. World food summit at Rome was held in

(*a*) 2000 (*b*) 1991

(*c*) 1981 (*d*) 1996

215. Panchayat Raj was first started in Rajasthan (1959) in the district

(*a*) Kota (*b*) Bikaner

(*c*) Nagour (*d*) Jodhpur

216. Lab to land programme was basically started for

(*a*) Small farmers (*b*) Marginal farmers

(*c*) Both (*a*) and (*b*) (*d*) None of these

217. Mandi ordinance of agriculture production was enforced in

(*a*) 1949 (*b*) 1959

(*c*) 1950 (*d*) 1955

218. Role of different agencies for village development is included in

(*a*) Resources mapping (*b*) Time line

(*c*) Transect (*d*) Chapati diagram

219. Gurgaon project was organised by

(*a*) S. Freud (*b*) F.L. Bryne

(*c*) J.P. Leagans (*d*) R.K. Meston

220. Sevagram attempt was started under the supervision of

(*a*) O.P. Dhama (*b*) H.W. Butt

(*c*) J.L. Nehru (*d*) M.K. Gandhi

221. Sriniketan attempt was started by R.N. Tagore in collaboration with Elmhirst in

(*a*) Jaipur (*b*) Bengal

(*c*) Coimbatore (*d*) Etawah

222. The latest development in participatory methodology among the following is

(*a*) Participatory Learning and Action

(*b*) Participatory Rural Appraisal

(*c*) Rapid Rural Appraisal

(*d*) None of these

223. The PRA technique wherein rural elders narrates life histories

(a) Ranking (b) Village transits

(c) Time line (d) None of these

224. Social calendar is a tool most commonly used in

(a) RRA (b) PRA

(c) LFA (d) Both (a) and (b)

225. IVLP stands for

(a) International Village Link Programme

(b) Institute Village Link Project

(c) Institute Village Link Programme

(d) Institute Village Linkage Programme

226. Who was popularized the bio-village concept ?

(a) Adam Smith (b) Devendra Thakur

(c) O.P. Dhama (d) M.S. Swaminathan

227. The ICAR Division of Extension was established in the year

(a) 1984 (b) 1971

(c) 1969 (d) 1955

228. The first transfer of technology project of ICAR was

(a) LLP (b) ORP

(c) KVK (d) AICPND

229. The KVK established as a result of

(a) B.R. Mehta committee (b) M.S. Mehta committee

(c) Chadha committee (d) M.S. Swaminathan committee

230. The first KVK was established in the year

(a) 1976 (b) 1975

(c) 1974 (d) 1973

231. The first KVK was established in

(a) Pondichery (b) Coimbatore

(c) Calcutta (d) Pantnagar

232. The demonstration in case of AICPND was done by

(a) Farmer (b) Media personal

(c) Extension person (d) Research scientist

233. Who was started Martandum attempt ?

(*a*) S. Hatch (*b*) A.H. Maslow

(*c*) Albert Mayer (*d*) F.L. Bryne

234. T and V system is a good example of

(*a*) Extension approach

(*b*) Training approach

(*c*) Co-operative self help approach

(*d*) Integrated development approach

235. Which is/are the main objective of the community development programme ?

(*a*) Area development

(*b*) Self help programme

(*c*) Development of the whole community

(*d*) All of these

236. The following scheme is exclusively meant for employment for rural youths is

(*a*) NREP (*b*) IRDP

(*c*) TRYSEM (*d*) MNREGA

237. Who was the leader of the Bhoodan movement ?

(*a*) O.P. Dhama (*b*) M.K. Gandhi

(*c*) Spencer Hatch (*d*) Acharya Vinoba Bhave

238. Which of the following organizations looks after the quality of Technical and Management education in India?

(*a*) UPCAR (*b*) ICAR

(*c*) CSIR (*d*) AICTE

239. National commission on farmer's was set up in India in

(*a*) March 2006 (*b*) February 2004

(*c*) June 2004 (*d*) March 2005

240. Who was inaugurated the National Agricultural Science Museum on 3rd Nov., 2004 ?

(*a*) Dr. Manmohan singh (*b*) A.B. Vajpayee

(*c*) Sharad Pawar (*d*) Dr. A.P.J. Abdul Kalam

241. National Agricultural Science Museum is situated at

(a) Hyderabad (b) Lucknow

(c) Kolkatta (d) New Delhi

242. ATMA stands for

(a) Assessment Transfer and Monitoring Agency

(b) Assessment Training and Management Agency

(c) Agricultural Technology and Management Agency

(d) None of these

243. ATMA is an important component under

(a) KVK (b) IISR

(c) NATP (d) NAEP

244. ATMA operates at

(a) Block level (b) State level

(c) District level (d) None of these

245. The ATMA under NATP is a

(a) Statutory body (b) Cooperative society

(c) Registered society (d) None of these

246. 'School on the air" is a project working in

(a) China (b) Mexico

(c) Srilanka (d) Philippines

247. The concept of Grameen Bank in Bangladesh is popularized by

(a) Aftab Hussain (b) Mohd. Younis

(c) Aga Khan (d) None of these

248. TRIPS stands for

(a) Trade Related Information Policy system

(b) Trade Related Intellectual Patent System

(c) Trade Related Intellectual Property Rights

(d) None of these

249. By eradicating poverty, who wished to "Wipe out every tear from every eye?

(a) Jawaharlal Nehru (b) Rabindra Nath Tagore

(c) Indira Gandhi (d) Mahatma Gandhi

250. Which is the basic unit of development under IRD programme ?

(a) A village (b) A family

(c) A community development block (d) State

251. National extension service (NES) in India was initiated on

(a) 2 October 1958 (b) 26 January 1956

(c) 26 January 1960 (d) 2 October 1953

252. First agriculture university in India was established at

(a) New Delhi (b) Hyderabad

(c) Coimbatore (d) Pantnagar

253. In the following identify one which is three -tier system of the local self goverment

(a) Gram Panchayat - Janpad Panchayat -Zila Panchayat

(b) Gram Panchayat - Village Co-operative -Khand Samiti

(c) IRD programme -TRYSEM -NREP

(d) Lab to land KVK-KGK

254. Level of communication are

(a) Conventional (b) Exploratory

(c) Participative (d) All the these

255. 'India Brand Equity Fund' was established in

(a) 1998 (b) 1996

(c) 1980 (d) 1990

256. Security and Exchange Board of India (SEBI) is a/an

(a) Advisory body (b) Constitutional body

(c) Statutory body (d) Autonomous body

257. A hypothesis is must in

(a) Experimental design (b) Explorative design

(c) Descriptive design (d) Random design

258. South Asian Association for Regional Co-operation (SAARC) was established on ?

(a) 8 December 1984 (b) 8 January 1984

(c) 8 December 1985 (d) 8 January 1985

259. What is the present number of member countries of European Economic Community (EEC).

(*a*) 12 (*b*) 6

(*c*) 21 (*d*) 36

260. India green revolution is the most successful in

(*a*) Wheat and potato (*b*) Jowar and oil seeds

(*c*) Wheat and rice (*d*) Milk and milk products

261. Economic planning is in

(*a*) Union list (*b*) State list

(*c*) Concurrent list (*d*) Both (*a*) and (*b*)

262. MRTP Acts is related to

(*a*) Monopoly and trade restrictions (*b*) Inflation control

(*c*) Transport control (*d*) None of these

263. Interest rate policy is a part of

(*a*) Fiscal policy (*b*) Industrial policy

(*c*) Monetary policy (*d*) None of the above

264. The basic of determining dearness allowance to employee in India is

(*a*) National income (*b*) Consumer price index

(*c*) Standard of living (*d*) Per capita income

265. Which day has been declared as 'Balika Diwas (Girl day) by the Ministry of Woman and Child Development ?

(*a*) 5 April, every year (*b*) 9 July, every year

(*c*) 5 October, every year (*d*) 9 December, every year

266. The toll free telephone number on Kisan call center is

(*a*) 5115 (*b*) 1551

(*c*) 1331 (*d*) 3113

267. The e- chaupal concept in extension delivery is promoted by

(*a*) MSSRF (*b*) ITC

(*c*) KINFRA (*d*) CAPART

268. ICAR was established based on the recommendation of the

(*a*) Ford foundation (*b*) World bank

(*c*) European Union (*d*) Royal commission

269. DARE was established in

(*a*) 1970 (*b*) 1980

(*c*) 1983 (*d*) 1973

270. MANAGE stands for

(*a*) National Institute of Agricultural Extension Management

(*b*) Management Institute for Agricultural Extension

(*c*) Institute of Agricultural Management

(*d*) None of these

271. In which year MANAGE was established in the year

(*a*) 1989 (*b*) 1984

(*c*) 1986 (*d*) 1986

272. The first agricultural university was established at

(*a*) Coimbatore (*b*) Kanpur

(*c*) Anand (*d*) Pantnagar

273. Post Graduate programme in extension was started in the year

(*a*) 1971 (*b*) 1955

(*c*) 1981 (*d*) 1964

274. The university education commission was headed by Dr. S. Radha Krishnan, recommended the establishment of

(*a*) Urban Universities (*b*) Rural Universities

(*c*) Both (*a*) and (*b*) (*d*) None of these

275. The rural system research idea was motivated by Dr. M.S. Swaminathan in

(*a*) 1981 (*b*) 1988

(*c*) 1977 (*d*) 2005

276. State Agriculture Universities in India were set up on the pattern of land grant college of

(*a*) China (*b*) France

(*c*) Mexico (*d*) U.S.A

277. According to CACP, sugarcane prices announced is

(*a*) Flat Prices (*b*) Procurement prices

(*c*) Statuary (*d*) None of these

278. Panchayat Raj came after

(*a*) Pre-determined programme (*b*) Self determined programme

(*c*) Fact determined programme (*d*) Both (*a*) and (*b*)

279. First step for making a programme planning includes

(*a*) Evaluation and teaching

(*b*) Analysis of the situation and determining problems

(*c*) Deciding on objectives

(*d*) None of these

280. Farm demonstration work began in ___ in USA by Dr. Seaman

(*a*) 1903 (*b*) 1950

(*c*) 1960 (*d*) 1935

281. The Royal commission's report came in

(*a*) 1928 (*b*) 1934

(*c*) 1945 (*d*) 1961

282. Government started new programme for social justice is/are

(*a*) SFDA (*b*) DPAP

(*c*) MFAL (*d*) All of these

283. A hypothesis is essentially a

(*a*) Interrogative statement (*b*) Relational preposition

(*c*) Empirical statement (*d*) General held view

284. 'Sagarmala' is a name associated with

(*a*) A project of port development (*b*) A drilling vessel

(*c*) Oil well in indian ocean (*d*) None of these

285. Pradhan Mantri Bharat Joda Pariyojana (renamed as NHDP) is related to ______

(*a*) Communication (*b*) Social integration

(*c*) Development of Sports (*d*) Development of highways

286. In national mineral policy (1933) which mineral was allowed for having investment from private sector

(*a*) Platinum (*b*) Iron

(*c*) Gold (*d*) None of these

287. Aam Admi Bima Yojana provides social security to ______

(*a*) All labours in rural areas

(*b*) All landless labours living below poverty line in rural areas

(*c*) All labours in urban areas

(*d*) All labours in both rural as well as urban areas

(*d*) Both (*a*) and (*c*)

288. Opinion polls fall under which one of the following design?

(*a*) Explorative design

(*b*) Descriptive design

(*c*) Diagnostic design

(*d*) Experimental design

289. Tarapore committee submitted its report on " full convertibility on rupees' in

(*a*) Current account

(*b*) Capital account

(*c*) Both (*a*) and (*b*)

(*d*) Special drawing rights

290. 'Sagarmala' has been renamed as

(*a*) National maritime development programme

(*b*) National highway development project

(*c*) National highway power project

(*d*) National Industrial Development Project

291. Value Added Tax (VAT) is imposed

(*a*) On first stage of production

(*b*) On final stage of production

(*c*) Directly on consumers

(*d*) On all stages between production and final sale

292. What do you mean by Mixed economy ?

(*a*) Co-existence of small and large industries

(*b*) Co-existence of public and private sectors

(*c*) Promoting both agriculture and industries in the economy

(*d*) Co-existence of rich and poor

293. The Community Development Project (1952) includes

(*a*) 8 pilot projects

(*b*) 15 pilot projects

(*c*) 28 pilot projects

(*d*) 32 pilot projects

294. Khadi and Village Industries Commission were established on

(a) 1980 (b) 1976

(c) 1969 (d) 1957

295. Chaudhary Charana Singh National Institute of Agricultural Marketing (NIAM) is situated at

(a) New Delhi (b) Jaipur

(c) Hisar (d) Hyderabad

296. National Agriculture Technology Project (NATP) was found from

(a) WHO (b) World Bank

(c) NAARM (d) ICAR

297. National food for work programme was launched on

(a) 14 November, 2004 (b) 2 October, 2004

(c) 14 November, 2000 (d) 2 October, 2000

298. The first chairman of the national commission on farmer's is

(a) Sompal (b) M.S. Swaminathan

(c) V.L. Chopra (d) K.L. Chadha

299. The members of Gram Sabha are ______

(a) Registered voters of Village Panchayat

(b) Sarpanch, Upsarpanch and Village level worker

(c) Sarpanch, Gram Sevak and elected Panchas

(d) None of these

300. Swarnajayanti Gram Swarojgar Yojna (SGSY) was launched by Government of India on:

(a) April, 1999 (b) April, 2007

(c) January, 2001 (d) March, 2002

301. Sampoorna Gramin Rojgar Yojana has been launched from

(a) 30th April, 2001 (b) 25th Sept, 2001

(c) 30th Sept, 2003 (d) None of these

302. Kutir Jyoti scheme is situated with ______

(a) Promoting cottage industry in villages

(b) Promoting employment among rural unemployed youth

(c) Providing electricity to rural families living Below the Poverty Line(BPL)

(d) All of these

303. CCEA(Cabinet committee on economic affairs) has decided to allow 'Navratna' and 'Mini Ratna companies to invest in mutual funds up-to percent of their cash balances.

(*a*) 15 per cent (*b*) 30 per cent

(*c*) 40 per cent (*d*) 50 per cent

304. The base of consumer price index for industrial workers is being shifted from 1982 to ______

(*a*) 2000 (*b*) 1998

(*c*) 1997 (*d*) 2001

305. South Asian University (SAU) is an international university sponsored by the eight member states of the South Asian Association for Regional Cooperation (SAARC) will have its head office in

(*a*) Dhaka (Bangladesh) (*b*) New Delhi (India)

(*c*) Colombo (Srilanka) (*d*) Male (Maldives)

306. Auto-Expo 2010 was organized in ______

(*a*) Mumbai (*b*) Hyderabad

(*c*) New Delhi (*d*) Bengaluru

307. Platinum jubilee year of RBI is ______

(*a*) 2009-10 (*b*) 2010-11

(*c*) 2011-12 (*d*) None of these

308. Which of the following is not the source of revenue of central government ?

(*a*) Excise duty (*b*) Income tax

(*c*) Agriculture income tax (*d*) Corporate tax

309. 'Pure banking nothing else' is a slogan raised by-

(*a*) ICICI (*b*) IDBI

(*c*) SBI (*d*) UTI bank

310. 'Smart Money' is a term used for

(*a*) Internet banking (*b*) Credit card

(*c*) Cash with bank (*d*) None of these

311. Which of the following has the maximum share in GSM mobile phone service market ?

(*a*) Vodafone (Earlier Hutch) (*b*) Airtel

(*c*) BSNL (*d*) Reliance

312. Which day is observed as National Women Literacy day?

(*a*) 8 September (*b*) 1 October

(*c*) 5 June (*d*) 8 February

313. The main objective of Training Rural Youth for Self Employment (TRYSEM) was

(*a*) To train rural youth for self employment

(*b*) To train urban youth for self employment

(*c*) To train urban and rural youth for self employment

(*d*) To provide finance for urban youth

314. The establishment of IORARC (India Ocean Rim Association for Regional Cooperation) was officially declared on

(*a*) 5 March 1977 (*b*) 5 March 1997

(*c*) 1April 1997 (*d*) 15 August 1947

315. Rural women can avail the benefit of Mahila Samriddhi Yojana if they open their account in

(*a*) Rural post offices (*b*) Commercial banks

(*c*) Rural development bank (*d*) Both (*b*) and (*c*)

316. Which of the following taxes have been removed in the union budget for 2009-10 ?

(*a*) FBT and CTT (*b*) Income tax

(*c*) GST (*d*) CENVAT

317. The cause of deflation is

(*a*) Lack of goods and services as compared to money supply

(*b*) Lack of import as compared to exports

(*c*) Lack of money supply as compared to supply of goods and services

(*d*) None of these

318. 'Aam Admi Bima Yojana' a social security scheme for rural landless households is implemented by the noddle agency ______

(*a*) National Insurance Co. (*b*) State government

(*c*) Oriental Insurance Co. (*d*) LIC

319. Who recommended minimum support price ?

(*a*) ICAR (*b*) State government

(*c*) Ministry of agriculture (*d*) CACP

320. The main mode of extension in the farmers first model

(a) Farmer to farmer
(b) Agent to farmer
(c) Researcher to agent
(d) None of these

321. Competition (Amendment) bill 2007 has replaced

(a) VAT
(b) MRTPC
(c) Electricity act, 2003
(d) None of these

322. 'Priyadarshini Project' is related to

(a) Empowerment of rural women
(b) Survival of girl child
(c) Free education to all girls
(d) Mid-day meal

323. Pratibha Kiran Yojana ' is a scheme to promote higher education among girls introduced by

(a) Karnataka
(b) Bihar
(c) Madhya Pradesh
(d) Rajasthan

324. 'Innovation lab' has been launched by

(a) Tata consultancy services
(b) Infosys technology
(c) Reliance
(d) Tata communications

325. The term evaluation was derived from

(a) English
(b) Roman
(c) Latin
(d) Greek

326. What is the last stage of programme planning process ?

(a) Evaluation
(b) Auditing
(c) Reconsideration
(d) Both (a) and (b)

327. Which technique applied to the projects that employ a fair and risk free technology ?

(a) PROMPT
(b) PRA
(c) PERT
(d) CPM

328. Which is the basic orientation of critical path method ?

(a) Deterministic
(b) Probabilistic
(c) Opportunistic
(d) None of these

329. Who was given the concept of PRA?

(*a*) Edger Dale (*b*) J.P. Leagans

(*c*) Nila Mukerjee (*d*) Henry Ford

330. Who is considered as father of PRA ?

(*a*) Torstan Hagestrand (*b*) Alexander Kapp

(*c*) Paul Leagons (*d*) Robert Chambers

331. The report on All India Rural Credit Enquiry Survey was submitted on

(*a*) 1961 (*b*) 1971

(*c*) 1954 (*d*) 1950

332. The term community development has originated in

(*a*) Oxford (*b*) China

(*c*) Cambridge (*d*) San Francisco

333. Community development project was started on the basis of

(*a*) Nalagarh Commission Report (*b*) Royal Commission Report

(*c*) GMF enquiry Report (*d*) None of these

334. "Extension Education in Community Development" is published by

(*a*) Ministry of Planning

(*b*) Ministry of Rural Development

(*c*) Ministry of Food and Agriculture

(*d*) Indian Council of Agricultural Research

335. The origin of the word democracy is from

(*a*) Roman (*b*) Greek

(*c*) Persian (*d*) None of these

336. What do you mean by the word democracy ?

(*a*) Peoples participation (*b*) Peoples government

(*c*) Peoples rule (*d*) Peoples judgment

337. The election to Panchayat is normally held after a period of

(*a*) 4 years (*b*) 5 years

(*c*) 2 years (*d*) 7 years

338. The basic unit of Panchayat Raj institutions

(*a*) Zila Parishad (*b*) Block Panchayat

(*c*) Panchayat Samithi (*d*) Grama Sabha

339. Who is the Secretary of Gram Panchayat ?

(a) Sarpanch (b) BDO
(c) Extension Officer (d) Gram Sevak

340. Ford foundation was invited to India on

(a) 1947 (b) 1958
(c) 1959 (d) 1960

341. Programme planning is a procedure of

(a) Working by the people (b) Working for the people
(c) Working with the people (d) All of these

342. The first KVK, Pondicherry was established in 1947 which related with

(a) IARI, New Delhi (b) TNAU, Coimbatore
(c) GBPAU&T, Pantnagar (d) ICAR, New Delhi

343. In which year Village Panchayat Act came into existence ?

(a) 1958 (b) 1961
(c) 1968 (d) 1975

344. Lab to programme was launched by the ICAR as a part of its ______ jubilee celebration

(a) Silver Jubilee (b) Golden Jubilee
(c) Platinum Jubilee (d) None of these

345. B. R. Mehta team stimulated an active consideration of ______ through democratic bodies in India

(a) Decentralization (b) Stabilization
(c) Centralization (d) Mobilization

346. ______ is a statement of situation objective problems and solutions

(a) A plan of work (b) Extension programme
(c) A calender of work (d) A evaluation

347. ______ is a plan of work arranged chronologically.

(a) Programme planning (b) A evaluation
(c) A plan of work (d) A calendar of work

348. The gap between the situation and objective is the area of ______

(a) Goal (b) Needs
(c) Interest (d) Both (a) and (c)

349. The government sponsored Firka development scheme was launched under the supervision/guidance of

(a) Luther Gullick
(b) J.L. Moreno
(c) O.P. Dhama
(d) T. Prakasam

350. In India ______ of the total population lives in village

(a) One third
(b) Half
(c) Two third
(d) Three fourths

351. National agriculture technology project (NATP) a World bank aided project was started in

(a) 1995 -96
(b) 1998-99
(c) 2000-01
(d) 2004-2005

352. Which is the largest source of national income in India ?

(a) Service sector
(b) Agriculture
(c) Industrial sector
(d) Trade sector

353. What do you mean by 'Public' sector ?

(a) Government ownership on commerce and trade
(b) Capitalist ownership on commerce and trade
(c) Private ownership on trade
(d) None of these

354. NABARD is a

(a) Bank
(b) Board
(c) Block
(d) Department

355. The term PRA was first used in

(a) USA
(b) India
(c) China
(d) Kenya

356. PRA was introduced in India by

(a) NGO'S
(b) ICAR
(c) Ford Foundation
(d) None of these

357. Which is the main innovation in PRA ?

(a) Methods
(b) Process
(c) Behaviour
(d) None of these

358. Which is the main innovation in RRA ?

(*a*) Process (*b*) Methods

(*c*) Behaviour (*d*) None of these

359. "Betting on the strong horse" is a slogan associated with which one of the following ?

(*a*) IRDP (*b*) HYVP

(*c*) SFMA (*d*) IADP

360. The IADP was popularly known as

(*a*) Training programme (*b*) Package programme

(*c*) Community programme (*d*) All of these

361. The term Green Revolution was coined by

(*a*) Norman E. Borlaoung (*b*) M.S. Swaminathan

(*c*) William Gadd (*d*) K.L. Chadha

362. The T and V system was sponsored by

(*a*) WHO (*b*) USAID

(*c*) ICAR (*d*) World Bank

363. The T and V system originated in

(*a*) India (*b*) Bangladesh

(*c*) China (*d*) Turkey

364. What is the main objective of NARP ?

(*a*) Research

(*b*) Extension

(*c*) Area and region specific research and extension in the state

(*d*) None of these

365. Encyclopedia of social work in India is published by

(*a*) ICAR (*b*) ICSSR

(*c*) UNESCO (*d*) Planning commission

366. In which year the first five year plan was submitted to the parliament ?

(*a*) 1953 (*b*) 1971

(*c*) 1951 (*d*) 1950

367. Harrod-Domar model was adopted in

(*a*) Sixth five-year plan (*b*) Fifth five -year plan

(*c*) Second five -year plan (*d*) First five- year plan

368. The term "Hindu rate of growth" was coined by

(*a*) J.N. Bhagawathi (*b*) Y.P. Singh

(*c*) B.R. Mehta (*d*) Raj Krishna

369. DRDA stands for

(*a*) District Rural Development Agency

(*b*) District Rural Development Authority

(*c*) District Rural Development Act

(*d*) None of these

370. DRDA is situated at

(*a*) District level (*b*) State level

(*c*) Block level (*d*) Village level

371. Council for Advancement of People's Action & Rural Technology (CAPART) is related with

(*a*) Assisting and evaluating rural welfare programmes

(*b*) Computer Software

(*c*) Consultant service of export promotion

(*d*) Controlling Water Pollution

372. In which year Securities and Exchange Board of India (SEBI) was established ?

(*a*) 1988 (*c*) 1989

(*b*) 1997 (*d*) 1971

373. The level of measurement attached to categorical variable is

(*a*) Nominal (*b*) Ordinal

(*c*) Interval (*d*) Ratios

374. In which year Operation Flood was launched ?

(*a*) 1971 (*b*) 1984

(*c*) 1961 (*d*) 1947

375. In which year Rashtriya Krishi Vikas Yojana was launched ?

(*a*) 2005 (*b*) 2007

(*c*) 1998 (*d*) 1987

376. In which year planning commission was set up ?

(*a*) 1961 (*b*) 1971

(*c*) 1950 (*d*) 1951

377. Who was the first chairman of Planning Commission

(a) Dr. Rajendra prasad
(b) Dr. S. Radakrishnan
(c) Indira Gandhi
(d) Pt. Jawaharlal Nehru

378. In which year TRYSEM was introduced ?

(a) 1969
(b) 1989
(c) 1979
(d) 1998

379. The Swarnajayanti Gram Swarozgar Yojna (SGSY) was launched on

(a) April 1987
(b) April 1999
(c) Arpril 1993
(d) April 2000

380. The share of center and state in the funding of SGSY is

(a) 50:50
(b) 40:60
(c) 60:40
(d) 75:25

381. Zamindari system was introduced by

(a) British
(b) Mughals
(c) Marathas
(d) None of these

382. Gurgaon attempt was in the state of

(a) Madhya Pradesh
(b) Uttar Pradesh
(c) Odisha
(d) Harayana

383. The concept of "Village guide" was introduced by;

(a) O.P. Dhama
(b) A.A. Reddy
(c) F.L. Brayne
(d) M K Gandhi

384. Who among the following is associated with YMCA

(a) L.H. Bailey
(b) Spencer Hatch
(c) S.K. Dey
(d) Carri Harrison

385. Royal commission on agriculture came out in the year

(a) 1937
(b) 1928
(c) 1910
(d) 1920

386. People's plan was envisaged by

(a) Jawaharlal Nehru
(b) M.K. Gandhi
(c) Mahalanobis
(d) M.N. Roy

387. NIRD under Ministry of Rural Development is located at

(a) Hyderabad (b) Lucknow

(c) Pantnager (d) New Delhi

388. AKIS stands for

(a) Agricultural Knowledge and Information Syndrome

(b) Agricultural Knowledge and Information Service

(c) Agricultural Knowledge and Information System

(d) None of these

389. The apex training institute at state level to give training support to ATMA

(a) MANAGE (b) CAPART

(c) SAMETI (d) SREP

390. According to Kelsey and Hearne the number of steps in programme planning is

(a) Three (b) Four

(c) Six (d) Seven

FILL IN THE BLANKS

1. Rural development at Sriniketan was initiated by_______
2. The general body of Panchayat is____________
3. ______ States have no Pachayat Raj
4. NATP is sponsored by______
5. The President of the ICAR is ________
6. National Institute for Transforming India Aayog was formed on _____
7. The first Krishi Vigyan Kendra (KVK) in India was setup at _________
8. The term 'Green Revolution" was coined by ______
9. Headquarters of CAPART are at______
10. The first Agriculture University in India is ________
11. Father of White Revolution in India is ________
12. Food and Agriculture Organization (FAO) headquarters at _________
13. "Rice Doctor" an interactive pest diagnostic expert system is developed by________
14. Kishan Credit Scheme, was launched in 1988-99 by_______

15. Boodhan Movement was started by ______
16. The Chairman of NITI Aayog is________
17. Jawahar Rozgar Yojana was started by _________
18. The Participatory Rural Appraisal (PRA) was developed at ______
19. The Govt. of India set up Planning Commission in ____________
20. The T and V system was proposed by___
21. The University Education Commission of 1949 which recommended the establishment of "Rural Universities" was headed by________
22. Planning Research and Action Institute (PRAI) is located at_____
23. "Krishi-Sudhar" was published from ________
24. The first institute in India to start teaching of extension education at undergraduate level was _________
25. The acronym ATMA stands for ________
26. The full form of ATIC is _________
27. The first Agri-clinic was set up at Vapi in ______
28. The "Garibi Hatao" slogan was given by ______
29. National Institute of Rural Developed (NIRD) under Ministry of Rural Development is located at __________
30. The Right to Information Act was passed in the year____

MARK TRUE OR FALSE

1. The impact of 'Green Revolution" was largely noted in Rice and wheat only.
2. World Food Summit at Rome was held in 1995.
3. Lab to land programme was started in the country as a part of ICAR Golden Jubilee Celebrations.
4. Nilokheri project was also known as Mazdoor Manzil.
5. National Development Council (NDC) is headed by Prime Minister.
6. TRIPS is related to an insect pest.
7. The Agriculture Price Commission (APC) was set in the year 1965.
8. The Participatory Rural Appraisal (PRA) was developed at Oxford University.
9. The headquarters of CAPART are at Hyderabad.
10. International Federation for Women in Agriculture has its headquarters at New Delhi.
11. Division of Extension was set up in India under the expert guidance of J.P. Leagans.

12. The first postgraduate programme in extension education was started in India in the year 1955.
13. The National Steering Committee (NSC) is Chaired by Director General, ICAR.
14. PRA was introduced in India by ICAR.
15. The Bio-village concept was popularized by Dr. M.S. Swaminathan.
16. Indian Society of Extension Education was established on 17th January 1965.
17. The KVK established as a result of M.S. Mehata Committee.
18. Zamindari System was introduced by Mughals.
19. The first Chairman of Planning Commission was Pt. Jawaharlal Nehru.
20. Free and compulsory education to all children in the age group of 6-14 years is envisaged in Kasturba Gandhi Balika Vidyalaya.
21. Kasturba Gandhi Balika Vidyalaya is meant for promoting education of girl children.
22. Summative evaluation is undertaken at the end of a programme.
23. The Directorate of Extension at the national level is located in the ministry of Human Resource Development.
24. "Operation Flood" programme relates to boosting milk supply.
25. Sriniketan project was started at Travancore.
26. The term "Community Development" appears to have originated from London.
27. The Chairman of State Development Committee is Chief Minister.
28. The National Service Scheme (NSS) was launched in the country in the year 1979.
29. Farm magazine "Gaon" was started by Bihar Government.
30. Verghese Kurien called the "Father of White Revolution in India.

ANSWERS

Multiple Choice Questions

1.(b)	2.(c)	3.(c)	4.(c)	5.(b)	6.(b)	7.(b)
8.(a)	9.(b)	10.(d)	11.(a)	12.(c)	13.(c)	14.(d)
15.(b)	16.(a)	17.(b)	18.(d)	19.(c)	20.(a)	21.(b)
22.(b)	23.(c)	24.(a)	25.(d)	26.(d)	27.(c)	28.(d)
29.(b)	30.(d)	31.(b)	32.(b)	33.(a)	34.(b)	35.(b)
36.(b)	37.(c)	38.(b)	39.(c)	40.(c)	41.(b)	42.(b)
43.(d)	44.(b)	45.(a)	46.(b)	47.(c)	48.(d)	49.(b)
50.(d)	51.(b)	52.(d)	53.(a)	54.(a)	55.(c)	56.(d)
57.(b)	58.(c)	59.(c)	60.(a)	61.(b)	62.(a)	63.(a)
64.(d)	65.(c)	66.(b)	67.(a)	68.(d)	69.(a)	70.(d)
71.(d)	72.(a)	73.(a)	74.(c)	75.(c)	76.(d)	77.(c)
78.(a)	79.(b)	80.(a)	81.(a)	82.(d)	83.(d)	84.(b)
85.(b)	86.(c)	87.(d)	88.(d)	89.(c)	90.(d)	91.(a)
92.(d)	93.(b)	94.(a)	95.(b)	96.(c)	97.(a)	98.(d)
99.(a)	100.(a)	101.(b)	102.(b)	103.(a)	104.(d)	105.(a)
106.(b)	107.(d)	108.(a)	109.(a)	110.(b)	111.(c)	112.(d)
113.(d)	114.(a)	115.(c)	116.(b)	117.(c)	118.(b)	119.(a)
120.(b)	121.(c)	122.(d)	123.(b)	124.(b)	125.(c)	126.(b)
127.(b)	128.(c)	129.(c)	130.(b)	131.(a)	132.(a)	133.(b)
134.(b)	135.(d)	136.(d)	137.(d)	138.(a)	139.(d)	140.(a)
141.(a)	142.(a)	143.(d)	144.(b)	145.(c)	146.(b)	147.(c)
148.(c)	149.(a)	150.(b)	151.(d)	152.(d)	153.(d)	154.(d)
155.(b)	156.(a)	157.(b)	158.(a)	159.(b)	160.(c)	161.(a)
162.(c)	163.(c)	164.(c)	165.(b)	166.(b)	167.(b)	168.(c)
169.(a)	170.(c)	171.(c)	172.(a)	173.(a)	174.(d)	175.(c)
176.(c)	177.(d)	178.(b)	179.(c)	180.(b)	181.(d)	182.(a)
183.(c)	184.(d)	185(d)	186.(b)	187.(a)	188.(c)	189.(b)
190.(a)	191.(d)	192.(a)	193.(b)	194.(a)	195.(b)	196.(a)
197.(b)	198.(a)	199.(d)	200.(b)	201.(b)	202.(d)	203.(b)
204.(d)	205.(c)	206.(d)	207.(d)	208.(c)	209.(c)	210.(b)
211.(a)	212.(b)	213.(a)	214.(d)	215.(c)	216.(b)	217.(a)
218.(d)	219.(b)	220.(d)	221.(b)	222.(a)	223.(c)	224.(b)
225.(d)	226.(d)	227.(b)	228.(d)	229.(b)	230.(c)	231.(a)
232.(d)	233.(a)	234.(b)	235.(d)	236.(c)	237.(d)	238.(d)
239.(b)	240.(d)	241.(d)	242.(c)	243.(c)	244.(c)	245.(c)
246.(d)	247.(b)	248.(c)	249.(d)	250.(b)	251.(d)	252.(d)
253.(a)	254.(d)	255.(b)	256.(c)	257.(a)	258.(c)	259.(a)
260.(c)	261.(c)	262.(a)	263.(c)	264.(b)	265.(d)	266.(b)
267.(b)	268.(d)	269.(d)	270.(a)	271.(d)	272.(d)	273.(b)
274.(b)	275.(b)	276.(d)	277.(c)	278.(c)	279.(b)	280.(a)

281.(a)	282.(d)	283.(b)	284.(a)	285.(d)	286.(c)	287.(b)
288.(b)	289.(b)	290.(a)	291.(d)	292.(b)	293.(b)	294.(d)
295.(a)	296.(b)	297.(a)	298.(a)	299.(a)	300.(a)	301.(b)
302.(c)	303.(b)	304.(d)	305.(b)	306.(c)	307.(a)	308.(c)
309.(c)	310.(b)	311.(b)	312.(a)	313.(a)	314.(b)	315.(a)
316.(a)	317.(c)	318.(b)	319.(d)	320.(a)	321.(b)	322.(a)
323.(c)	324.(a)	325.(c)	326.(c)	327.(d)	328.(a)	329.(d)
330.(d)	331.(c)	332.(c)	333.(c)	334.(c)	335.(b)	336.(c)
337.(b)	338.(d)	339.(d)	340.(c)	341.(c)	342.(b)	343.(a)
344.(b)	345.(a)	346.(b)	347.(d)	348.(b)	349.(d)	350.(d)
351.(b)	352.(a)	353.(a)	354.(a)	355.(d)	356.(a)	357.(c)
358.(b)	359.(d)	360.(b)	361.(c)	362.(d)	363.(d)	364.(c)
365.(d)	366.(c)	367.(d)	368.(d)	369.(a)	370.(a)	371.(a)
372.(a)	373.(a)	374.(a)	375.(b)	376.(c)	377.(d)	378.(c)
379.(b)	380.(d)	381.(b)	382.(b)	383.(c)	384.(b)	385.(b)
386.(d)	387.(a)	388.(c)	389.(c)	390.(d)		

Fill in the Blanks

1. R N Tagore
2. Gram Sabha
3. Meghalaya
4. World Bank
5. Agriculture Minister
6. 1st Jan, 2015
7. TNAU&T, Coimbatore
8. William Gadd
9. New Delhi
10. GBPUAT, Pantnagar
11. Verghese Kurien
12. Rome
13. IRRI
14. SBI
15. Acharya Vinoba Bhave
16. Prime Minister
17. Rajeev Gandhi
18. Clark Unversity
19. 1950
20. Daniel Benor
21. Dr S. Radhakrishanan
22. Lucknow
23. Agra
24. College of Agriculture, Calcutta University
25. Agricultural Technology Management Agency
26. Agricultural Technology Information Center
27. Bulsar
28. Indira Gandhi
29. Hyderabad
30. 2005

Mark True or False

1. True
2. False
3. True
4. True
5. True
6. False
7. True
8. False
9. False
10. True
11. True
12. True
13. True
14. False
15. True
16. False
17. True
18. True
19. True
20. False
21. True
22. True
23. False
24. True
25. True
26. False
27. True
28. False
29. True
30. True

CHAPTER 4

MANAGEMENT IN AGRICULTURAL EXTENSION

Multiple Choice Questions

1. **The word 'Management' was derived from**
   (a) French (b) English
   (c) Latin (d) Greek

2. **For AON notation, ... is used to show the task itself**
   (a) Circle (b) a Box
   (c) Triangle (d) Rectangle

3. **In networking techniques, ... show the sequence in which work is done**
   (a) Arrow (b) Line
   (c) Double line (d) Triple line

4. **In AoA notation, the .... represents an event - either the beginning of another activity or completion of previous one.**
   (a) Node (b) Triangle
   (c) Circle (d) Line

5. **Activities on the critical path having zero slack / float time**
   (a) Critical path (b) Critical Activity
   (c) Critical circle (d) None of these

6. **The literal meaning of the word 'Management' is**
   (a) Shop keeping (b) House keeping
   (c) Store keeping (d) Staff placement

7. **'POSCORB' as management function was developed by**
   (*a*) Urwick (*b*) L. Gullick
   (*c*) Henry Fayol (*d*) Gullick and Urwick

8. **Planning Research and Action Institute is situated at**
   (*a*) New Delhi (*b*) Mumbai
   (*c*) Bangaluru (*d*) Lucknow

9. **The ability of an individual to induce or influence the beliefs or action of other persons of group.**
   (*a*) Leadership (*b*) Authority
   (*c*) Power (*d*) Control

10. **The right in a position to exercise discretion in making decisions affecting others**
    (*a*) Responsibility (*b*) Hierarchy
    (*c*) Unity of command (*d*) Authority

11. **EMT–lab is associated with**
    (*a*) Rural development (*b*) Agricultural development
    (*c*) Entrepreneurship development (*d*) Water management

12. **The concept of POSDCORB in management was proposed by**
    (*a*) Karl Marx (*b*) Emile Durkheim
    (*c*) Gullick (*d*) Henry Fayol

13. **The "acid test " in administration is related to**
    (*a*) Arrangement of Hierarchy (*b*) Public relation
    (*c*) Supervision (*d*) None of these

14. **Which of the following is the correct equation?**
    (*a*) Motivation + Empowerment= Enabling
    (*b*) Enabling + Empowerment + Motivation
    (*c*) Motivation + Enabling = Empowerment
    (*d*) None of these

15. **The term " entrepreneur" was coined by**
    (*a*) Tony Pears (*b*) William Gadd
    (*c*) Richard Cantinlon (*d*) None of these

16. A recognizable work item of a project requiring time and resource for its completion.

(*a*) Activity (*b*) Event

(*c*) Slack (*d*) None of these

17. An activity that indicates precedence relationship and requires no time nor resource.

(*a*) Copy activity (*b*) Double activity

(*c*) Dummy Activity (*d*) None of these

18. An instantaneous point in time signifying completion or beginning of an activity.

(*a*) Activity (*b*) Event

(*c*) Dummy Activity (*d*) None of these

19. The total time on this path is the shortest duration of the project.

(*a*) Critical Path (*b*) Critical activity

(*c*) Critical event (*d*) Critical time

20. The weighted average of the estimated optimistic, most likely and pessimistic time duration of a project activity:

(*a*) Expected Zone (*b*) Expected Event

(*c*) Expected Time (*d*) Expected Activity

21. POSDCoRB is developed by

(*a*) Gullick (*b*) Drucker

(*c*) Fayol (*d*) Desai

22. Process of placing right man on right job is called

(*a*) Coordinating (*b*) Placement

(*c*) Directing (*d*) Leading

23. Proponent of participative management

(*a*) Likert (*b*) McGregor

(*c*) Tony Pears (*d*) William Godd

24. Under path goal approach of management, leadership is categorised into .... groups

(*a*) Two (*b*) Five

(*c*) Four (*d*) Eight

25. Management By Objectives was developed by

(a) Simon and March
(b) Likert
(c) Davis
(d) Peter Drucker

26. Official examination of records is called as

(a) Report
(b) Audit
(c) Budget
(d) None of the above

27. The "Theory of human motivation " is given by

(a) C.R. Desai
(b) Henry Fayol
(c) A.H. Maslow
(d) M.N. Davison

28. The term " bureaucracy" has been coined by

(a) P.V. Peresion
(b) Henry Fayol
(c) E.M. Swamson
(d) Mark Weber

29. The term" credibility has been given by

(a) Argyris
(b) Hovland
(c) Alderfer
(d) Taylor

30. Refers to the ability of an individual to adapt his behaviour to the demands of the situation.

(a) Self checking
(b) Self-esteem
(c) Self evaluation
(d) Self-monitoring

31. Refers to the degree of liking an individual has for himself.

(a) Self-esteem
(b) Self-evaluation
(c) Self checking
(d) None of these

32. .... Refers to the individual's ability to withstand stress

(a) Emotional development
(b) Emotional stability
(c) Emotional clarity
(d) Emotional evaluation

33. Evaluating an employee's current and or past performance relative to his or her performance standards.

(a) Performance Appraisal
(b) Performance calculation
(c) Performance evaluation
(d) None of these

34. The procedure for determining the duties and skill requirements of a job and the kind of person who should be hired for it.

(a) Job profile
(b) Job analysis
(c) Job checking
(d) Job Context

35. A systematic comparison done in order to determine the worth of one job relative to another.

(*a*) Job evaluation (*b*) Job profile

(*c*) Job checking (*d*) Job content

36. The last stage in the process of recruitment is

(*a*) Appointment (*b*) Placement

(*c*) **Orientation** (*d*) Designing

37. The essence of the scalar principle is

(*a*) Span of control (*b*) Unity of command

(*c*) Communication (*d*) Coordination

38. Brainstorming is basically a

(*a*) Motivation technique (*b*) Creative technique

(*c*) Evaluation technique (*d*) Reasoning technique

39. The term "entrepreneur" is coined by

(*a*) Tony Pears (*b*) Maslow

(*c*) Richard Cantinlon (*d*) Gilford Pinchot

40. The most important for an entrepreneur to have is

(*a*) High knowledge about the business (*b*) High need for achievement

(*c*) High operational skills (*d*) None of these

41. Human skills and conceptual skills are more significant.

(*a*) Lower level (*b*) Higher level

(*c*) Middle-management level (*d*) Uncertain

42. First-level managers require more in order to supervise operational employees.

(*a*) Communication skills (*b*) Technical skills

(*c*) Manual skills (*d*) None of these

43. It refers to the participation of employees in the decision- making process.

(*a*) Communication (*b*) Centralization

(*c*) Decentralization (*d*) None of these

44. It refers to the retention of control by the top management in the area of decision making.

(*a*) Decentralization (*b*) Centralization

(*c*) Authority (*d*) Communication

45. It is the process by which managers allocate a chunk of their workload to their subordinates.

(a) Leadership (b) Communication
(c) Delegation (d) Authority

46. The behaviour domain that assumes significance in lecture presentation

(a) Effective (b) Cognitive
(c) Connative (d) Both (a) and (c)

47. T- group training is grouped under

(a) Transactional method (b) Experiential method
(c) Sensitivity method (d) None of these

48. The co-trainer concept of training is very much related to

(a) L group (b) O group
(c) Q group (d) T group

49. The stage of brainstorming in which free flow of ideas take place

(a) Orange stage (b) Green stage
(c) Red stage (d) White stage

50. Evaluation of ideas in brainstorming takes place in

(a) White stage (b) Red stage
(c) Green stage (d) Yellow stage

51. The method of verbal presentation on a topic by a speaker to group of audience

(a) Group discussion (b) Lecture
(c) Panel discussion (d) Symposium

52. Discussion papers based on in-depth study and research is presented and discussed under the guidance of experts

(a) Conference (b) Buzz session
(c) Seminar (d) Workshop

53. Span of control deals with

(a) Power given to an authority
(b) Responsibilities given to an authority
(c) The number of persons that one can supervise
(d) None of these

54. The Abbreviation POSDCORB has been denoted by

(*a*) Max Weber (*b*) Luther Gullick

(*c*) E.W. Riggs (*d*) Rober K. Kalz

55. The number of subordinates an officer can effectively supervise is known as

(*a*) Hierarchy (*b*) Span of control

(*c*) Authority (*d*) None of these

56. The '.....................'is a "tendency to let the assessment of an individual one trait influence the evaluation of that person on other specific traits."

(*a*) Attribution error (*b*) Halo effect

(*c*) Ranking method (*d*) None of these

57. Owen recommended the use of a "...................." to openly rate employee's work on a daily basis.

(*a*) Silent content (*b*) Silent worker

(*c*) Silent evaluator (*d*) Silent monitor

58. HRM objectives are................................. Societal, organizational, functional and personal.

(*a*) Four fold (*b*) Six fold

(*c*) Three fold (*d*) One fold

59. The Principles of Scientific Management in 1911 is written by

(*a*) Taylor (*b*) Gullick

(*c*) Adam Smith (*d*) Fayol

60. .... is the process of generating a sufficiently large group of applicants from which to select qualified individuals for available jobs.

(*a*) Directing (*b*) Leading

(*c*) Recruiting (*d*) Coordinating

61. It is also referred as manpower planning, personnel planning or employment planning

(*a*) Human resource planning (*b*) Human resource development

(*c*) Human resource matrix (*d*) Human resource management

62. National Institute of Entrepreneurship and Small Business Development is located at

(*a*) Nagpur (*b*) Lucknow

(*c*) Jaipur (*d*) New Delhi

63. The office of district collector of India is based on the principle of

(a) Delegation (b) Decentralization
(c) De-concentration (d) None of these

64. The Indian Institute of Public Administration is located at

(a) Jodhpur (b) Lucknow
(c) Kanpur (d) New Delhi

65. The learning curve in training follows a

(a) C-shape (b) S-shape
(c) U -shape (d) None of these

66. The first step in SATNA need assessment method is

(a) Problem identification (b) Job description
(c) People identification (d) None of these

67. "Training for Organizational Transformation" is a book authored by

(a) Pareek, Rao and Parimjee (b) Berger and Berger
(c) Lynton and Pareek (d) Henry Fayol

72. Keeping all the activities with in the specified limits so that the organization is able to attain its objectives through proper utilization available resources, facilities, manpower and time is

(a) Planning (b) Coordinating
(c) Controlling (d) None of these

73. Allocation of funds for different sectors of programme

(a) Recruiting (b) Budgeting
(c) Financing (d) None of these

74. A new concept of budgeting as recommended by Administrative Reforms Commission is

(a) Peripheral budgeting (b) Performed budgeting
(c) Performance budgeting (d) None of these

75. Official examination of records is

(a) Evaluation (b) Audit
(c) Budget (d) All of these

76. International Conference on Extension Strategy for minimizing risk in rain fed agriculture was held in 1991 at

(a) Jodhpur (b) Hyderabad
(c) Mumbai (d) New Delhi

77. Father of Development Administration ?

(*a*) George Gnatt (*b*) Henry Fayol

(*c*) Max Weber (*d*) P.V. Peresion

78. Theory X and Theory Y was developed by

(*a*) Davis (*b*) McGregor

(*c*) Alderfer (*d*) Emile Durkheim

79. Hierarchy of Need theory was proposed by

(*a*) Hovland (*b*) Berlo

(*c*) Maslow (*d*) Stellings

80. Need theory was proposed by

(*a*) McClelland (*b*) Kurt Lewin

(*c*) Simon and March (*d*) Davis

81. The principles of scientific management developed by

(*a*) C.R. Desai (*b*) A.H. Maslow

(*c*) Frederick W. Taylor (*d*) M.N. Davison

82. Conceptual skills are usually not very essential for the managers at the supervisory level it is

(*a*) True (*b*) Can not say

(*c*) False (*d*) Option is not available

83. Functional authority violates the principle of.

(*a*) Leading (*b*) Direction

(*c*) Unity of command (*d*) None of the above

84. Father of scientific management is

(*a*) Taylor (*b*) Rober K. Kalz

(*c*) Fayol (*d*) None of the above

85. .... are those which are directly responsible for accomplishing the objectives of the enterprise, while staff functions are advisory in nature.

(*a*) Power functions (*b*) Supervisory functions

(*c*) Line functions (*d*) None of the above

86. Curriculum literally means

(*a*) Content in Training (*b*) Keep on running

(*c*) Theory in training (*d*) All of these

87. The S in the SMART objective refers to

(a) Severe (b) Simple

(c) Sincere (d) Specific

88. First step in SQ3R method is

(a) Study (b) Select

(c) Survey (d) Calculation

89. The first stage in ELC cycle is

(a) Publishing (b) Experiencing

(c) Generalizing (d) None of these

90. The fulcrum of ELC cycle is

(a) Processing (b) Experiencing

(c) Generalizing (d) Specializing

91. The costs associated with these times are called respectively the .... and the ....

(a) Crash time and Normal time (b) Normal cost and Crash cost

(c) Normal time and crash activity (d) Crash-cost and Normal time

92. Bar chart as a tool of project monitoring was given by

(a) W. Quarter (b) K. Venkatswarulu

(c) H. L. Gantt (d) T.S. Osborne

93. Who is the father of scientific management ?

(a) M. Weber (b) Henry L. Gantt

(c) L. Urwick (d) Fredrick Taylor

94. An informal communication type involved in the spread of rumor

(a) Horizontal communication (b) Vertical communication

(c) Diagonal communication (d) Grapevine communication

95. Frank Gilbreth is considered the "......................"

(a) Father of Scientific Study (b) Father of Motion Study

(c) Father of Human Relation (d) Father of Behavioural Study

96. Brainstorming method was given by

(a) Taylor (b) Fayol

(c) Osgood (d) Weber

97. The number of participants is minimum in

(a) Conference discussion (b) Huddle method
(c) Symposium method (d) Buzz session

98. T group training with trainers from different discipline

(a) Internal group (b) Formal group
(c) Stranger group (d) None of these

99. Buzz session is normally conducted for

(a) 60-120 minutes (b) 30-40 minutes
(c) 20-30 minutes (d) 5-10 minutes

100. Training programme for VLWs in Training and Visit system is organized on a

(a) Weekly basis (b) Monthly basis
(c) Fortnightly basis (d) Yearly basis

101. Project management techniques can be classified under two broad categoeis i.e., Bar Charts and Networks that is

(a) True (b) Can not say
(c) False (d) None of these

102. Gantt Charts developed by

(a) Henry Fayol (b) F. Taylor
(c) Henry L Gantt (d) L. Gullick

103. It is a pictorial representation specifying the short and finish trifle for various tasks to be performed in a project on a horizontal time-scale.

(a) Pie-diagram (b) Bar chart
(c) Big chart (d) Gantt Chart

104. A recognizable work item of a project requiring time and resource for its completion.

(a) Event (b) Circle
(c) Activity (d) Results

105. Slack or float is used to indicate the spare time available within a non-critical activity.

(a) False (b) True
(c) Can not say (d) None of the above

106. The concept of microlab was first used in

(*a*) India (*b*) USA

(*c*) Germany (*d*) China

107. The rule of control of higher over the lower is known as

(*a*) Hierarchy (*b*) Power

(*c*) Administration (*d*) Responsibility

108. The term management is derived from

(*a*) Latin (*b*) English

(*c*) French (*d*) Greek

109. National Agricultural Research and Extension Coordination Committee meets

(*a*) Once in a year (*b*) Twice in a year

(*c*) Thrice a year (*d*) Uncertain

110. Zonal Research and Extension Action Committee (ZREAC) meets

(*a*) Once a year (*b*) Twice a year

(*c*) Thrice a year (*d*) Uncertain

111. Each ZREAC meeting runs for

(*a*) One day (*b*) Two days

(*c*) Four days (*d*) Six days

112. Each District Agricultural Research and Extension Implementation Committee meeting runs for

(*a*) One day (*b*) Two days

(*c*) Three days (*d*) Uncertain

113. An expression of discontent or dissatisfaction with any aspect of organization is

(*a*) Stress (*b*) Conflict

(*c*) Grievance (*d*) None of these

114. The democratic or participative style of leadership became popular during the era of

(*a*) Behaviour theory (*b*) Human relation theory

(*c*) Systems theory (*d*) Ecological theory

115. "Intelligence Quotient" is calculated by using the formula

(a) Chronological age X Mental age/100

(b) Mental age/Chronological age X 100

(c) Chronological age/Mental age X 100

(d) None of these

116. The concept of "Intelligence Quotient" was proposed by

(a) William Stung (b) Rober K, Kalz

(c) Henry Fayol (d) M. Weber

117. The term "motivation" is derived from ______

(a) Latin Language (b) German Language

(c) French Language (d) Greek Language

118. ERG theory was developed by

(a) Tony Pears (b) William Gadd

(c) J S Adams (d) C. Alderfer

119. Equity theory was given by

(a) J S Adams (b) Maslow

(c) Simon and March (d) Fayol

120. Valence theory was developed by

(a) Maslow (b) M.N. Davison

(c) Karl Mark (d) V H Vroom

121. Reinforcement theory was given by

(a) Skinner (b) Alderfer

(c) Hovland (d) Taylor

122. The word 'Management' is derived from

(a) Latin (b) German

(c) Roman (d) French

123. The word 'Management' is derived from which word

(a) Menage (b) Menege

(c) Manage (d) None of these

124. The normal pace of delivery in the lecture method is

(a) 260 words per minute (b) 110 words per minute

(c) 160 words per minute (d) 180 words per minute

125. The apex body to offer training for extension functionaries is

(*a*) IIHR (*b*) IISR

(*c*) NAARM (*d*) MANAGE

126. Training by feeling is

(*a*) Reflective observation (*b*) Abstract conceptualization

(*c*) Concrete experience (*d*) Active experimentation

127. The role of trainer in experiential approach is

(*a*) Controller (*b*) Contributor

(*c*) Facilitator (*d*) Evaluator

128. What is the last phase in experiential learning cycle ?

(*a*) Processing (*b*) Application

(*c*) Publication (*d*) Experience

129. Theory can be integrated in the experiential learning cycle at

(*a*) Processing (*b*) Specializing

(*c*) Applying (*d*) Generalizing

130. Proponent of participative management is the

(*a*) Thurstone (*b*) Adam Smith

(*c*) Likert (*d*) Gullick

131. Under path-goal approach of leadership, the leadership behaviour is categorized into

(*a*) Two Groups (*b*) Four Groups

(*c*) Four Groups (*d*) Six Groups

132. The downward communication in an organization flows from

(*a*) Subordinates to superior (*b*) Superior to subordinate

(*c*) Subordinates to subordinate (*d*) None of these

133. The upward communication in an organization flows from

(*a*) **Subordinates to superior** (*b*) Superior to subordinate

(*c*) Subordinates to subordinate (*d*) None of these

134. Feedback information from subordinates to superior flows through

(*a*) **Upward communication** (*b*) Downward communication

(*c*) Horizontal communication (*d*) None of these

135. The communication that takes place between two or more functionaries at the same level under the same superior

(*a*) Upward communication (*b*) Downward communication

(*c*) Horizontal communication (*d*) None of these

136. The time for the activity at minimum cost is called .... and the minimum time for the activity is called ....

(*a*) Normal time and Crash time (*b*) Normal time and crash activity

(*c*) Crash time and Normal time (*d*) None of the above

137. Time-and -motion" study method introduced by

(*a*) Winslow (*b*) C. Alderfer

(*c*) Taylor (*d*) Weber

138. Who was the most prominent of the administrative theorists ?

(*a*) F.W. Riggs (*b*) March and Simon

(*c*) Max Weber (*d*) Henry Fayol

139. According to whom, "a bureaucracy is a highly structured, formalized and impersonal organization."

(*a*) Carl Marx (*b*) Smith and Jones

(*c*) Weber (*d*) Taylor

140. Mary Parker Follet advocated ....

(*a*) Power sharing (*b*) Power delegation

(*c*) Power communication (*d*) Managerial roles approach

141. 'Father of the Human Relations Approach"

(*a*) Elton Mayo (*b*) C. Alderfer

(*c*) Maslow (*d*) Weber

142. .... gave maturity immaturity theory.

(*a*) Gullick (*b*) Maslow

(*c*) Alderfer (*d*) Chris Angyris

143. Who define management as "the process of designing and maintaining an environment in which individuals, working together in groups, efficiently accomplish selected aims."

(*a*) Weber and Weber (*b*) Harold Koontz and Heinz Weihrich

(*c*) Adam and Smith (*d*) None of the above

144. The functions of a manager provide a useful framework for organizing management knowledge under the various heads of planning, organizing, staffing, leading and controlling.

(*a*) False (*b*) True

(*c*) Can not say (*d*) None of the above

145. It is a normal practice to categorize management into three basic levels: (1) top-level management, (2) middle level management, and (3) supervisory of first-level management.

(*a*) True (*b*) Uncertain

(*c*) False (*d*) None of these

146. Henry Mintzberg devised a new approach .... by observing what managers actually do.

(*a*) Authority approach (*b*) Managerial roles approach

(*c*) Power communication (*d*) Power delegation

147. The need for technical skills is lesser at the.

(*a*) Top level (*b*) Medium level

(*c*) Higher level (*d*) Uncertain

148. Using a method called structured observation, Mintzberg isolated .... roles.

(*a*) Five (*b*) Zero

(*c*) Twenty (*d*) Ten

149. Who outlined a new theory called Theory Z ?

(*a*) Willaim Gadd (*b*) Good and Hatt

(*c*) William Ouchi (*d*) None of the above

150. Pilot Project on Technology Assessment and Refinement was started by Division of Agricultural Extension, ICAR in the year

(*a*) 1990 (*b*) 1995

(*c*) 1998 (*d*) 1980

151. Father of Development administration is the

(*a*) George Gant (*b*) Max Weber

(*c*) T.S. Osborne (*d*) M.L. Johansson

152. The concept of unity of command is complementary to the principle of

(*a*) Supervision (*b*) Scalar Chain

(*c*) Authority (*d*) None of these

153. The theory of Maslow as the "major wrong theories was given by

(*a*) Donald Schewab (*b*) Riggs

(*c*) Michael Nash (*d*) None of these

154. Systems approach is also known as

(*a*) Systems theory (*b*) Integrated theory

(*c*) Modern theory (*d*) Classical theory

155. The term "motivation" comes from Latin roots "movere to mean

(*a*) To bring forward (*b*) To move or incite into action

(*c*) To bring one against wishes (*d*) None of these

156. It refers to a manager granting the right to a subordinate to make decisions or use his discretion in judging certain issues.

(*a*) Unity of command (*b*) Delegation of authority

(*c*) Unity of direction (*d*) None of these

157. It is understood as the process of forecasting an organization's future demand for, and supply of the right type of people in the right numbers.

(*a*) Human Resource Management (*b*) Human Resource Designing

(*c*) Human Resource Matrix (*d*) Human Resource Planning

158. Refers to the extent to which an individual is prepared to take risks.

(*a*) Risk bearing (*b*) Risk calculation

(*c*) Risk-taking (*d*) Risk evaluation

159. Centralization is the retention of decision-making authority with the top management, whereas decentralization is granting of decision-making powers to the lower-level employees.

(*a*) No (*b*) Can not say

(*c*) Yes (*d*) None of these

160. .... is an activity undertaken with a specific objective involving a specific time frame with a definite beginning and a definite completion.

(*a*) Project (*b*) Activity

(*c*) Plan (*d*) Idea

161. Project Life cycle has .... of Conception, Definition, Execution and operation, which every project passes through various phases of life cycles synonym to life cycles of living being.

(*a*) Class (*b*) Category

(*c*) Phases (*d*) Styles

162. The .... is the process of critical examination and analysis of the proposal in totality.

(a) Project definition
(b) Project implementation
(c) Project appraisal
(d) Project expenditure

163. It is a pattern of relationships through which people undertake the directions of managers to pursue the common goals for the organization.

(a) Organization
(b) Management
(c) Leadership
(d) None of these

164. It indicates a set of expected behavior patterns attributed to someone occupying a given position in an organization.

(a) Role
(b) Role clarity
(c) Role disparity
(d) None of these

165. A distinct area of management practices to meet the challenges of rapidly growing economies.

(a) Project development
(b) Project planning
(c) Project Management
(d) Project matrix

166. .... are the pictorial representation of various tasks required to be performed for accomplishment of the project objectives.

(a) Bar diagrams
(b) Bar charts
(c) Milestone charts
(d) Pie-diagram

167. Which is the process of helping people to acquire competences ?

(a) Human Resource Management
(b) Human Resource Development
(c) Human Resource Planning
(d) None of the above

168. The event which occurs only when more than one activities are accomplished.

(a) Critical event
(b) Merge Event
(c) Burst Event
(d) None of the above

169. The use of a piece-rate incentive system advocated by

(a) Taylor
(b) William
(c) E.W. Riggs
(d) Karl Marx

170. Who identified three kinds of skills for administrators ?

(a) Max Weber
(b) Rober K. Kalz
(c) Rogers and Rogers
(d) Henry Fayol

FILL IN THE BLANKS

1. The theory of human motivation is given by ________
2. Latest theory of motivation is __________
3. The term 'Entrepreneur' is coined by ______
4. Management grid was developed by______
5. The idea of scientific management was introduced by ________
6. The term 'Bureaucracy' has been given by _______
7. The concept of micro-lab was first used in _________
8. 'Art of leadership' is written by ___________
9. The term 'Theory X" was coined by _______
10. "Brain Storming" method was developed by ________
11. Father of development administration is ________
12. The term "Credibility" was coined by ______
13. Time and Motion study method introduced by_____
14. Frank Gilbreth is considered the_______
15. The father of scientific management is ______
16. Gantt Charts developed by _______
17. According to whom a bureaucracy is a highly structured, formalized and impersonal organization ________
18. Three kinds of skill for administrators identified by ________
19. The use of piece rate incentive system advocated by _____
20. Process of placing right man on right job is called ________

MARK TRUE OR FALSE

1. Mary Parker Follect advocated Power Sharing.
2. The most prominent of the administrative theorists are Taylor.
3. Slack or Float is used to indicate the spare time available within a non-critical activity.
4. Leniency is also called as strictness tendency or constant errors.
5. Valence theory was given by Alderfer.
6. Management by objectives was developed by Peter Drucker.

7. Hierarchy of Need theory was given by Maslow.
8. Elton Mayo is the father of the Human Relations Approach.
9. The need for technical skills is less at the medium level.
10. William Ouchi outlined a new theory known as Theory Z.
11. Mary Parker Follet advocated Power Sharing.
12. Official examination of records is known as budget.
13. Functional authority violates the principles of unity of command.
14. E.W. Riggs identified three kinds of skill.
15. The learning curve in training follows a S-shape.
16. The term 'entrepreneur' is coined by Richard Continlin.
17. The office of district collector of India is based on the principle of delegation.
18. The Indian Institute of Public Administration is located at New Delhi.
19. Brainstorming is basically reasoning techniques.
20. MANAGE is the apex body to offer training for extension functionaries.

ANSWERS

Multiple Choice Questions

1.(a)	2.(b)	3.(a)	4.(c)	5.(b)	6.(b)	7.(b)
8.(d)	9.(c)	10.(d)	11.(c)	12.(c)	13.(c)	14.(c)
15.(c)	16.(a)	17.(c)	18.(b)	19.(a)	20.(c)	21.(a)
22.(b)	23.(a)	24.(c)	25.(d)	26.(b)	27.(c)	28.(c)
29.(b)	30.(d)	31.(a)	32.(b)	33.(a)	34.(b)	35.(a)
36.(c)	37.(b)	38.(b)	39.(c)	40.(b)	41.(c)	42.(b)
43.(c)	44.(b)	45.(c)	46.(b)	47.(c)	48.(d)	49.(c)
50.(c)	51.(b)	52.(c)	53.(c)	54.(b)	55.(b)	56.(b)
57.(d)	58.(a)	59.(a)	60.(c)	61.(a)	62.(d)	63.(c)
64.(d)	65.(b)	66.(a)	67.(c)	68.(b)	69.(b)	70.(c)
71.(b)	72.(c)	73.(b)	74.(c)	75.(b)	76.(d)	77.(a)
78.(b)	79.(c)	80.(a)	81.(c)	82.(a)	83.(c)	84.(a)
85.(c)	86.(b)	87.(b)	88.(c)	89.(b)	90.(a)	91.(b)
92.(c)	93.(a)	94.(d)	95.(b)	96.(c)	97.(d)	98.(c)
99.(d)	100.(c)	101.(a)	102.(c)	103.(d)	104.(c)	105.(b)
106.(b)	107.(a)	108.(c)	109.(b)	110.(c)	111.(b)	112.(a)
113.(c)	114.(b)	115.(b)	116.(a)	117.(a)	118.(d)	119.(a)
120.(d)	121.(a)	122.(d)	123.(a)	124.(a)	125.(d)	126.(c)
127.(c)	128.(b)	129.(d)	130.(c)	131.(c)	132.(b)	133.(a)
134.(a)	135.(c)	136.(a)	137.(c)	138.(d)	139.(c)	140.(a)

141.(a)	142.(d)	143.(b)	144.(b)	145.(a)	146.(b)	147.(c)
148.(d)	149.(c)	150.(b)	151.(a)	152.(b)	153.(c)	154.(c)
155.(b)	156.(b)	157.(d)	158.(c)	159.(c)	160.(a)	161.(c)
162.(c)	163.(a)	164.(c)	165.(b)	166.(b)	167.(b)	168.(b)
169.(a)	170.(b)					

Fill in the Blanks

1. A.H. Maslow
2. ERG Theory
3. Richard Cantinlon
4. Blake and Mouton
5. Taylor
6. Marx Weber
7. USA
8. Ordinary Tead
9. Mc Gregor
10. Osgood
11. George Gant
12. Hovland
13. Taylor
14. Father of Motion Theory
15. M. Weber
16. Henry L. Gantt
17. Weber
18. Rober K. Karl
19. Taylor
20. Placement

Mark True or False

1. True
2. False
3. True
4. True
5. False
6. True
7. True
8. True
9. False
10. True
11. True
12. False
13. True
14. False
15. True
16. True
17. False
18. True
19. False
20. True

CHAPTER 5

TRAINING AND RESEARCH METHODOLOGY

Multiple Choice Questions

1. **Which is the experimental method of data collection ?**

   (*a*) Field experiment (*b*) Sample survey

   (*c*) Field study (*d*) None of these

2. **Questionnaire involves presentation of**

   (*a*) Oral written stimuli (*b*) Oral verbal stimuli

   (*c*) Written verbal stimuli (*d*) None of these

3. **Ranking of raw data in numerical order is known as**

   (*a*) Scores (*b*) Series

   (*c*) Points (*d*) Frequency distribution

4. **Which of the following accounts for existing data, predict new observations and guide further research:**

   (*a*) Theory (*b*) Method

   (*c*) Hypothesis (*d*) Practical

5. **Research design is not evidentially a**

   (*a*) Structure (*b*) Plan

   (*c*) Strategy (*d*) Product

6. **What is the basic function of a good research design?**

   (*a*) Minimize error variance (*b*) Decrease experimental variance

   (*c*) Brief metaphysical observation (*d*) Maximise extra error variance

7. **To study the effect of independent variable on a single experiment we go through**

   (*a*) RBD | (*b*) LSD
   (*c*) CRD | (*d*) Factorial design

8. **Opinion polls falls under**

   (*a*) Experimental design | (*b*) Descriptive design
   (*c*) Explorative design | (*d*) Diagnostic design

9. **Which is the example of discrete variable?**

   (*a*) Sex | (*b*) Weight
   (*c*) Height | (*d*) Age

10. **Most people retain ______ per cent of what they read.**

    (*a*) 5-10 | (*b*) 10-15
    (*c*) 25-30 | (*d*) 35-40

11. **Symposium is a short series of lectures; usually by ______**

    (*a*) 1 -2 speakers | (*b*) 2 - 5 speakers
    (*c*) 5 - 7 speakers | (*d*) 10 - 15 speakers

12. **Philips 66 format or hurdle system is related with**

    (*a*) Panel | (*b*) Form
    (*c*) Buzz sessions | (*d*) Conference

13. **Best method for selection of leader is**

    (*a*) Sociometry | (*b*) Election
    (*c*) Discussion method | (*d*) Selection

14. **Age is classified under which of the following variables**

    (*a*) Continuous variable | (*b*) Intervening variable
    (*c*) Discrete variable | (*d*) None of these

15. **A hypothesis when tested true become a**

    (*a*) Theory | (*b*) Principle
    (*c*) Philosophy | (*d*) Law

16. **The bridge between theory and empirical enquiry is connected through**

    (*a*) Research hypothesis | (*b*) Research principle
    (*c*) Research problem | (*d*) Research output

17. A tentative explanation of an event or observation

(*a*) Hypothesis (*b*) Principle
(*c*) Variable (*d*) Argument

18. Survey method would most typically used in

(*a*) Clinical assessment (*b*) Laboratory experiment
(*c*) Opinion poll (*d*) All of these

19. The commonly used method in a social survey is

(*a*) Observation (*b*) Questionnaire
(*c*) Interview (*d*) Case study

20. Assignment of values to alternatives is known as

(*a*) Observation (*b*) Evaluation
(*c*) Judgment (*d*) Analysis

21. Probability of a leap year will have 53 Sundays is

(*a*) 52/53 (*b*) 2/53
(*c*) 2/7 (*d*) 4/7

22. The modified version of a panel discussion in which three or more resource persons discuss a specific topic is known as

(*a*) Colloquium (*b*) Symposium
(*c*) Workshop (*d*) Conference

23. The most preferred approach in extension work is

(*a*) Autocratic (*b*) Democratic
(*c*) Lessiz -faire (*d*) None of these

24. The central element in an effective learning situation is

(*a*) Teacher (*b*) Learner
(*c*) Subject matter (*d*) Both (*a*) and (*b*)

25. "Cone of experience" was devised by

(*a*) Kelsey and Hearne (*b*) Edgar Dale
(*c*) Seaman A. Knapp (*d*) Alexander Knapp

26. A conjectural statement of relationship between the variables is called

(*a*) Research objective (*b*) Research theme
(*c*) Research problem (*d*) Research hypothesis

27. "To control variance" is the major purpose of

(a) Statistical test (b) Research design

(c) *Ex-post-facto* method of research (d) Interview schedule

28. Who believes that the work will be done if you leave workers on their own

(a) Autocratic leader (b) Democratic leader

(c) *Laissez-faire leader* (d) Lay leader

29. The art and science of teaching adults is known as

(a) Adultology (b) Andragogy

(c) Agogology (d) Topology

30. Cross-section of population is represented less in

(a) Simple random sample (b) Multistage sample

(c) Cluster sample (d) Stratified random sample

31. A special type of working conference, usually of a week or more in duration with lectures, individual conferences giving emphasis on small working group is known as

(a) Conference (b) Panel discussion

(c) Symposium (d) Workshop

32. The most effective as well as most expensive method of extension work is

(a) Demonstration (b) Farm and home visit

(c) Meetings (d) Lecture

33. The idea of testing hypothesis was set forth by

(a) S. Tyson (b) J. Neyman (1928)

(c) A. Wald (d) E.L. Lehman

34. In 1933, the theory of testing of hypothesis was propounded by

(a) Karl Pearson (b) R.A. Fisher

(c) E.L. Lehman (d) T. Tyson

35. The hypothesis under test is

(a) Simple hypothesis (b) Alternate hypothesis

(c) Null hypothesis (d) None of the above

36. Hypothesis that states the negation of the assertion is commonly called as a

(a) Null hypothesis (b) Alternate hypothesis

(c) Parallel hypothesis (d) Logical hypothesis

37. Scientific data must be

(a) Objective (b) Measurable

(c) Predictable (d) All of these

38. Interview is a widely used technique in

(a) Local study (b) Public study

(c) Empirical studies (d) Non-scientific studies

39. Whether a test is one tailed or two tailed depends on

(a) Null hypothesis (b) Alternate hypothesis

(c) Parallel hypothesis (d) Logical hypothesis

40. The best control for practice effect in an experiment is

(a) Random selection (b) Counter balancing

(c) Equivalent group (d) Split half presentation

41. To study average rate of change in population,we use

(a) Median (b) Mode

(c) Mean (d) GM

42. Most elementary level of measurement is

(a) Nominal (b) Ordinal

(c) Interval (d) All of these

43. Cross section of population is represented less in

(a) Simple random sampling (b) Multistage sampling

(c) Cluster sampling (d) None of these

44. Measurement in social science is related to

(a) Objects (b) Properties of objects

(c) Phenomenon (d) None of these

45. Steps in teaching are shortly called as

(a) AIETA (b) AIDACS

(c) AIDCAS (d) KPDC

46. The India an Council of Social Science Research is situated at

(a) New Delhi (b) Mumbai

(c) Lucknow (d) Bikaner

47. The book " Scientific Social Survey and Research" is written by

(*a*) C R Kothari (*b*) P C Tripathi

(*c*) P V Young (*d*) R P Singh

48. "After this, therefore caused by this " is usually denoted to indicate

(*a*) Experimental design (*b*) Survey design

(*c*) Explorative design (*d*) *Ex post facto* design

49. The term *ex post facto* was first coined by

(*a*) Festinger and Katz (*b*) Chapin and Greenwood

(*c*) Goode and Hatt (*d*) None of these

50. A hypothesis is a must in

(*a*) Explorative design (*b*) Descriptive design

(*c*) Experimental design (*d*) Both (*a*) and (*b*)

51. In Sequential Probability Ratio Test (SPRT),the sample size is

(*a*) Fixed (*b*) Fixed but small

(*c*) Fixed but large (*d*) A random variable

52. In SPRT, decision about hypothesis Ho is taken

(*a*) After each successive observation (*b*) At least after five observations

(*c*) After a fixed number of observations (*d*) When the experiment is over

53. To decide about SPRT involves

(*a*) One region only (*b*) Two regions only

(*c*) Three regions (*d*) Multi-regions

54. SPRT was started by

(*a*) R.A. Fisher (*b*) A. Wald

(*c*) G.W. Snedecer (*d*) Thomas Bayes

55. The decision criterion in SPRT depends on the function of

(*a*) Type I error (*b*) Type II error

(*c*) Type I and II errors (*d*) None of these

56. "System approach to training" has been given by

(*a*) R. A. Fisher (*b*) B. Richard

(*c*) Alfred York (*d*) Sameul Paul

57. The difference between " the expected and actual job performance " is known as

(a) Training content
(b) Training assessment
(c) Training need
(d) All of these

58. The process of transforming the experience into knowledge as a basic for action is called

(a) Continued learning
(b) Topical learning
(c) Experimental learning
(d) Experiential learning

59. The term "Pedagogy" is a

(a) Latin term
(b) French term
(c) Roman term
(d) Greek term

60. Experiential learning is

(a) Instructor centered
(b) Learner centered
(c) Subject matter centered
(d) All of these

61. The spiral model of training has been coined by

(a) E. Hagen
(b) Kingson and Davis
(c) Swanson and Brown
(d) Lynton and Pareek

62. "NOIR" is associated with which one of the follwing ?

(a) Reliability
(b) Measurement
(c) Validity
(d) Certification

63. A systematic body of knowledge is called

(a) Theory
(b) Technology
(c) Science
(d) Arts

64. The book entitled "Art of leadership" is written by

(a) C.W.Merrifield
(b) Ordway Tead
(c) Alvin Gouldner
(d) D. Berlo

65. The book entitled 'Leadership in voluntary organizations is written by

(a) Rogers Bellow
(b) C.W. Merrifield
(c) H.B. Trecker
(d) A.A. Livereiht

66. The father of statistics is

(a) C.R. Maniologa
(b) D.N. Fisher
(c) R.A. Fisher
(d) James Rots Tam

67. The farther of learning is

(*a*) Plato (*b*) Aristotle

(*c*) Knapp (*d*) Thorndike

68. A research paper is a brief report of research work based on

(*a*) Primary Data only (*b*) Secondary Data only

(*c*) Both Primary and Secondary Data (*d*) None of these

69. Newton gave three basic laws of motion. This research is categorized as

(*a*) Descriptive Research (*b*) Fundamental Research

(*c*) Sample survey (*d*) Applied Research

70. In the process of conducting research "Formulation of Hypothesis" is followed by

(*a*) Statement of Objectives (*b*) Collection of Data

(*c*) Survey of Data (*d*) Analysis of Data

71. What do you means by *Ex Post Facto* research ?

(*a*) The research is carried out prior to the incident

(*b*) The research is carried out after the incident

(*c*) The research is carried out along with the happening of an incident

(*d*) All of these

72. Portion of population is known as

(*a*) Representative (*b*) Population

(*c*) Sample (*d*) All of these

73. What is the key of accurate sampling ?

(*a*) Opinion (*b*) Prediction

(*c*) Randomness (*d*) None of these

74. Parge is not a reliable measure of

(*a*) Dispersion (*b*) Mode

(*c*) Mean (*d*) Median

75. Who is the father of statistics?

(*a*) Carl Pearson (*b*) Emile Duke man

(*c*) Carl Pearson (*d*) Francis Galton

76. The measure of dispersion which ignores signs of deviation from the central value

(a) Quartile deviation (b) Standard deviation
(c) Mean deviation (d) None of these

77. Which measure of dispersion is least affected by extreme values

(a) Mean deviation (b) Standard deviation
(c) Quartile deviation (d) None of these

78. Empirical relation between quartile deviation and standard deviation is represented as

(a) 4QD= 4SD (b) 3QD= 2SD
(c) 6QD= 5SD (d) 4QD= 4SD

79. The relation between standard deviation and mean deviation is

(a) 2MD=3SD (b) 3MD= 2SD
(c) 5MD= 4SD (d) 6MD=5SD

80. The average of sum of squares of deviation about the mean is known as

(a) Absolute deviation (b) Mean deviation
(c) Standard deviation (d) Variance

81. The simplest way of visualizing a correlation is to construct a

(a) Graph (b) Scatter diagram
(c) Table (d) Bar diagram

82. Regression coefficient takes values

(a) $-\infty$ to 1 (b) -1 to ∞
(c) -1 to $+1$ (d) ∞ to $+\infty$

83. If correlation coefficient is 0.7, then coefficient of determination is

(a) 0.49 (b) 0.70
(c) 0.07 (d) 0.60

84. Measure of dispersion includes the range and

(a) Mode (b) Median
(c) Standard deviation (d) Mean

85. The range of multiple correlation coefficient R is

(a) -1 to $+1$ (b) 0 to ∞
(c) 0 to 1 (d) None of these

86. Equality of two regression coefficient can be tested with

(*a*) ANOVA (*b*) 't' test

(*c*) Z test (*d*) χ^2 test

87. The range of X^2 statistic is

(*a*) –1 to –1 (*b*) – α to +α

(*c*) 0 to α (*d*) – α to 1

88. The range of statistic t is

(*a*) – α to + α (*b*) –1 to –1

(*c*) 0 to 1 (*d*) 0 to – α

89. The range of variable ratio F is

(*a*) 0 to α (*b*) 1 to + α

(*c*) –1 to –1 (*d*) – 0 to – α

90. The first step in summarizing the data is

(*a*) Classification (*b*) Tabulation

(*c*) Graphical representation (*d*) Both (*a*) and (*b*)

91. The method of extension in which small-group discussion is used to divide a large of persons into subgroups for greater participation

(*a*) Discussion 44 (*b*) Discussion 77

(*c*) Discussion 66 (*d*) None of these

92. The pioneer of intrinsic programming in learning is

(*a*) Kikert (*b*) Marvin

(*c*) Chambers (*d*) Crowder

93. Vocational training is one of the most important functions of the integrated model of

(*a*) MANAGE (*b*) ATIC

(*c*) ATMA (*d*) KVK

94. A measuring device is ______ if the results obtained from it are in close approximation of what the true results are

(*a*) **Valid** (*b*) Reliable

(*c*) Practical (*d*) Incorrect

95. A series of steps or stages in which both tools and methods are used to contribute is

(*a*) Progress (*b*) Profess

(*c*) Process (*d*) None of these

96. The term "pedagogy" is derived from ______

(*a*) Greek language (*b*) German language

(*c*) Arabic language (*d*) Latin language

97. Student's t-test was given by

(*a*) Davis Thomson (*b*) W.S. Gosset

(*c*) W.G. Cochran (*d*) G.W. Snedecor

98. Student's t-test is application in case of

(*a*) Small samples (*b*) Large samples

(*c*) For samples size 5-30 (*d*) None of the above

99. Bar Chart was developed by

(*a*) R.A. Fisher (*b*) Horburt Luskenner

(*c*) Hennery L. Gantt (*d*) Pricidy Trigrid

100. The general micro-unit of an agro-ecosystem is

(*a*) District (*b*) Household

(*c*) Village (*d*) Block

101. The TRIM in communication implies

(*a*) Technology, Recording, Initiating, Meaning

(*b*) Target, Receiver, Impact, Method

(*c*) Training, Replication, Information, Monitoring

(*d*) Technology, Reinforcement, Improvement, Manipulation

102. The book entitled "Psychometric theory " is authored by

(*a*) A.F. Zanderr (*b*) K.C. Pratt

(*c*) Jum C. Nunnally (*d*) F.N. Kerlinger

103. "Socio-metric technique " was coined by

(*a*) G. Murphy (*b*) M. Karlson

(*c*) H.P. Edwin (*d*) J.L. Moreno

104. The book entitled " Training for development " is written by

(*a*) Lynton and Pareek (*b*) B.P. Sinha

(*c*) P.V. Young (*d*) R.A. Fischer

105. "Semential differential scale" has been given by

(*a*) Karlson (*b*) Likert

(*c*) Chauve (*d*) Osgood

106. The book entitled "The measurement of meaning" is a book written by

(a) D.C. Mishra (b) S.K. Wagmare

(c) H.P. Edwin (d) Charles E. Osgood

107. The book entitled "New dimensions in extension training" is written by

(a) R.P. Singh (b) A.K. Mishra

(c) A.H. Maslow (d) D.C. Mishra

108. The true evaluation of the students is possible by

(a) Evaluation at the end of the course

(b) Evaluation twice in a year

(c) Formative evaluation

(d) Continuous evaluation

109. A hypothesis is a

(a) Law (b) Postulate

(c) Supposition (d) Act

110. Controlled group condition is applied in

(a) Survey Research (b) Historical Research

(c) Descriptive Research (d) Experimental Research

111. Workshops are meant for

(a) Giving lectures (b) Multiple target groups

(c) Hand on training/experience (d) Show new theories

112. In the following research tool is

(a) Graph (b) Questionnaire

(c) Illustration (d) Table

113. Jury opinion is used to establish

(a) Adjustability (b) Creativity

(c) Validity (d) None of these

114. Index of discrimination is otherwise known as

(a) Item credibility index (b) Item validity index

(c) Item difficulty index (d) None of these

115. The word logic have its root in

(a) French (b) German

(c) Latin (d) Greek

116. Logic is a science of

(*a*) Reality (*b*) Truth

(*c*) Foam (*d*) Both (*a*) and (*b*)

117. The first step in summarizing the data is known as

(*a*) Calculation (*b*) Interpretation

(*c*) Tabulation (*d*) Classification

118. The first step in analysis of data is known as

(*a*) Manipulation (*b*) Interpretation

(*c*) Categorization (*d*) Summarization

119. The value of co-efficient of concordance range between

(*a*) -1 to +1 (*b*) 0 + 1

(*c*) -1 to 0 (*d*) $-\alpha$ to $+\alpha$

120. Formula for calculating coefficient of variation is

(*a*) Mean/SD x SD (*b*) SD/Mean x 100

(*c*) Mean x SD/100 (*d*) None of these

121. Correlation coefficient is independent of

(*a*) Scale (*b*) Origin

(*c*) Neither (*a*) and (*b*) (*d*) Both (*a*) and (*b*)

122. The range of simple correlation coefficient is

(*a*) -1 to +1 (*b*) $1 + \alpha$

(*c*) 0 to 1 (*d*) $-\alpha$ to $+\alpha$

123. Which is the unit of correlation coefficient ?

(*a*) No unit (*b*) kg/cc

(*c*) Percent (*d*) ppm

124. The measure of association of relationship between two or more variables is known as

(*a*) Dispersion (*b*) Regression

(*c*) Correlation (*d*) Deviation

125. The value of x^2 always lies between

(*a*) -1 to $+1$ (*b*) 0 to 1

(*c*) 0 to ∞ (*d*) $-\infty$ to 1

126. Who is considered as father of statistics ?

(a) Emile Durkheim (b) Ronald Fisher

(c) Francis Galton (d) Berlo CK

127. If the measure of Kurtosis is greater than 3, then the curve is

(a) Normal (b) Platykurtic

(c) Mesokurtic (d) Leptokurtic

128. The value of probability ranges from

(a) -1 to $+1$ (b) 0 to 1

(c) 0 to ∞ (d) $-\infty$ to $+1$

129. Probability of two persons being born on the same day is

(a) 1/7 (b) 1/49

(c) 1/365 (d) 1/27

130. The training objectives must be SMART. The SMART means

(a) Specific, Measurable, Attainable, Realistic, Time bound

(b) Simple, Manageable, Attainable, Realistic, Temporary

(c) Similar, Manageable, Attributed, Rational, Time bound

(d) None of these

131. The word "Curriculum" is derived from

(a) French (b) German

(c) Latin (d) English

132. The concept of "Micro-lab" in training was first of all used in

(a) UK (b) USA

(c) Germany (d) China

133. The optimum duration for conducting micro-lab is

(a) 35-45 minutes (b) 1-1.5 Hour

(c) 300-3.5 Hours (d) 3.5 - 4.5 Hours

134. The originator of "Linear programming" in "programmed learning" is

(a) Skinner (b) Knowles

(c) Thurstone (d) Daniel

135. Brainstorming method was originated in

(a) USA (b) UK

(c) Germany (d) China

136. Adult learning is

(*a*) Subject center (*b*) **Problem centered**

(*c*) Institute centered (*d*) Phase centered

137. The term' '*andragogy* " is derived from the Greek terms " *ander*" meaning _____ and "*agogos*" meaning _____ respectively

(*a*) Adult and following (*b*) Adult and reading

(*c*) Adult and leading (*d*) None of these

138. The concept of semantic differentiation was given by

(*a*) Fisher (*b*) Mereno

(*c*) Osgood (*d*) E M Rogers

139. Q Methodology was put forwarded by

(*a*) Fisher (*b*) Mereno

(*c*) Osgood (*d*) D. Berlo

140. The three R's in the SQ3R strategy of comprehension does not include

(*a*) Read (*b*) Review

(*c*) Reward (*d*) None of these

141. Questionnaire involves presentation of

(*a*) Written-verbal stimuli (*b*) Oral-verbal stimuli

(*c*) Oral-written stimuli (*d*) Both (*a*) and (*b*)

142. Degree of freedom of X^2 test for 4x3 contingency table is

(*a*) 4 (*b*) 6

(*c*) 9 (*d*) 8

143. Mean Deviation is minimum when taken about

(*a*) Mean (*b*) Mode

(*c*) Median (*d*) None of these

144. The relation between arithmetic mean, geometric mean and harmonic mean is represented as

(*a*) AM=GM=HM (*b*) AM<GM<HM

(*c*) AM> GM> HM (*d*) None of these

145. The most frequent score in a given distribution

(*a*) Mean (*b*) Median

(*c*) Mode (*d*) Both (*a*) and (*b*)

146. Most commonly used method in a social survey is

(a) Observation (b) Interview

(c) Questionnaire (d) Case study

147. The collection of data in research falls under

(a) Statistical design (b) Observational design

(c) Operational design (d) Sampling design

148. Survey method would most typically used in

(a) Clinical assessment (b) Opinion poll

(c) Laboratory experiment (d) Field experiment

149. Statistical conclusion are always

(a) Absolute true (b) Absolutely wrong

(c) True on average (d) None of these

150. The process of drawing conclusion or inferences after a careful analysis of collected data

(a) Interpretation (b) Tabulation

(c) Classification (d) Demonstration

151. The best measure of dispersion among the following is

(a) Standard deviation (b) Mean deviation

(c) Coefficient of deviation (d) Median deviation

152. "A priori" literally means

(a) Agree with reason (b) Agree without reason

(c) Both (a) and (b) (d) None of these

153. The basic difference between classical research and participatory research lies in their

(a) Content (b) Control

(c) Power (d) Time

154. Social Science Index is published from

(a) Netherlands (b) India

(c) Columbia (d) New York

155. The International Index of Social Science named Social Science Citation Index is published from

(a) Netherlands (b) USA

(c) USA (d) India

156. Directory of Social Science Research Institutions in India is published by

(a) NIRD (b) ICSSR

(c) UNESCO (d) ICAR

157. The cone of experience was developed by

(a) Karl Pearson (b) Egder Dale

(c) Kuldeep Nair (d) SC Parmer

158. The term Homophily and Heterophily were given by

(a) Rogers (b) Lazersfield and Merton

(c) Gabriel Trade (d) Francis Galton

159. The theory of social change was put forwarded by

(a) K Lewin (b) William Stephenson

(c) E M Rogers (d) P Leaganes

160. Unfreezing existing behaviour and motivating people to change is expressed as

(a) Interest (b) Desire

(c) Conviction (d) None of these

161. The learning curve in teaching follows a

(a) O-shape (b) S-shape

(c) U-shape (d) None of these

162. Education is shaped by

(a) Knowledge (b) Teacher

(c) Student (d) Society

163. For testing the significance of correlation coefficient, we use:

(a) t-test (b) ANOVA

(c) χ^2 test (d) F-test

164. The use of two negatives in the same sentence is known as:

(a) Repression (b) Paralogic

(c) Equivocation (d) Negation

165. Information gained through direct observation and measurement is known as:

(a) Observational evidence (b) Empirical evidence

(c) Experimental variance (d) None of these

166. Planning Research and Action institute is located at:

(a) Mumbai
(b) New Delhi
(c) Nagpur
(d) Lucknow

167. Which is the correct chronology of study method suggested by A. Comte?

(a) Experiment, observation, comparison and history
(b) Observation, experiment, history and comparison
(c) History, observation, experiment and comparison
(d) Experiment, observation, history and comparison

168. Which is the hindrance in objective finding?

(a) Respondent is aware of the magnitude of the problem
(b) Investigator is unbiased
(c) Investigator is serious in investigating a problem
(d) None of these

169. Which procedures cannot prove useful for ensuring objective fact finding?

(a) Trained people should be set to the field
(b) Responded should be given monetary incentives
(c) Senior people should be sent th the field
(d) Some counter sample checking should be done.

170. Which is the dependable method for obtaining objectivity?

(a) Field survey
(b) Group discussion
(c) Questionnaire method
(d) Interview method

171. Which is the primary methodology of functionalists?

(a) Field work
(b) Positivism
(c) Library work
(d) Conjectural history

172. For collecting information about Indian society method preferred is:

(a) Social survey method
(b) Interview method
(c) Questionnaire method
(d) Participant observation method

173. In the following, which methods is usually not used for objective fact findings?

(a) Library method
(b) Historical method
(c) Questionnaire method
(d) None of these

174. An observation, uninfluenced by personal bias is known as

(a) Critical (b) Subjective

(c) Functional (d) Objective

175. Objectivity can't be introduced at the time of

(a) Meeting non-respondents (b) Field survey

(c) Data collection (d) Data interpretation

176. Observation means to see things

(a) Without a purpose (b) With a purpose

(c) For important discussion (d) With many purposes

177. Generally case study aims to

(a) Expose persons danger to society

(b) Establish statistical correlation

(c) Bring out the structure of the unit as a whole

(d) None of these

178. Crucial feature of scientific data is its

(a) Quality (b) Reliability

(c) Universal applicability (d) All of these

179. An important criterion of a good sample is its

(a) Large size (b) Representativeness

(c) Small size (d) None of these

180. Altitude is in nature and character

(a) Clear (b) Vague

(c) Objective (d) Subjective

181. Objective finding is hindered because

(a) Respondents are usually biased

(b) Respondents are usually illiterate

(c) Investigation are educated

(d) None of these

182. Things external to the mind, material objects and overt bodily behaviour are

(a) Chemical events (b) Physical events

(c) Subjective events (d) Objective events

183. Schedule is a list of questions which will be answered in an interview by a/an

(*a*) Respondent (*b*) Interviewer

(*c*) Researchers (*d*) Surveyor

184. Questionnaire method is not applicable if

(*a*) Area is vast (*b*) There is no overpopulation

(*c*) There is wide spread literacy (*d*) There is illiteracy

185. Case study involves:

(*a*) Very careful and complete observation of a person

(*b*) Complete observation of a person

(*c*) Careful observation of a person

(*d*) None of these

186. Thurston scale is a type of

(*a*) Questionnaire (*b*) Interview schedule

(*c*) Altitude scale (*d*) All of these

187. Training Performance Index does not includes

(*a*) Training utility index (*b*) Training cost index

(*c*) Training participation index (*d*) None of these

188. A Science is to be judged by there criteria, reliability of its knowledge, its organization and :

(*a*) Its identifiability

(*b*) Its method

(*c*) Bulk of qualitative data it can produce

(*d*) Its capacity to expose universal laws

189. Which is an important feature of a good sample?

(*a*) Big quantity (*b*) Correct representativeness

(*c*) Smaller quantity (*d*) None of these

190. Some times mailed questionnaire is preferable to interview because

(*a*) Face to face contact is possible (*b*) People immediately return it

(*c*) It works out cheaper (*d*) None of these

191. Which method is less expensive in comparison to other method?

(*a*) Mailing Questionnaire (*b*) Interview

(*c*) Case study (*d*) Questionnaire

192. The basic purpose of an interview is to assess the candidate's

(*a*) Power of expression
(*b*) General knowledge
(*c*) Intellectual capacity
(*d*) Personal Qualities

193. Purpose of time study is

(*a*) To give timely assistance
(*b*) To determine fair day's work
(*c*) To remove wastage of time
(*d*) All of these

194. Cyclic model of training was developed by

(*a*) R.D. Miller
(*b*) A. York
(*c*) D.G. Reddy
(*d*) Lynton and Pareek

195. System approach of training is basically a:

(*a*) Experimental approach
(*b*) Productive approach
(*c*) Diagnostic approach
(*d*) Descriptive approach

196. Training principle that best explain the use of analogue is :

(*a*) Readiness
(*b*) Association
(*c*) Intensity
(*d*) Capacity

197. The training stage that does not from a component of cyclic model:

(*a*) Pre-training
(*b*) Preparatory training
(*c*) Post-training
(*d*) Training

198. The training approach development from engineering concept

(*a*) Process approach
(*b*) Performance approach
(*c*) Discrete approach
(*d*) System approach

199. The last phase in experimental learning cycle is

(*a*) Generalization
(*b*) Application
(*c*) Experience
(*d*) Publication

200. The role trainer in experimental approaches is

(*a*) Controller
(*b*) Enabler
(*c*) Contributor
(*d*) Facilitator

FILL IN THE BLANKS

1. The term *ex post facto* was first coined by __________
2. Q-methodology was put forwarded by ________
3. The book entitled "Sampling Technique" is written by ___________
4. The book entitled "Scientific Social Survey and Research" is written by ____
5. The concept of semantic differentiation was discovered by __________
6. Freedom from variable errors is known as ________
7. Logic is a science of _________
8. The first step in summarizing the data is known as ________
9. An interval in which scores are grouped is known as__________
10. The first step in analysis of data is __________
11. A hypothesis is must in ______
12. Book that covers anyone respect but in a detailed manner is known as__________
13. The Indian Council of Social Science Research is located at _______
14. A conjectural statement of relationship between the variable is called _________
15. Sociometric method first time used by __________

MARK TRUE OR FALSE

1. F-test is used to test ratio of two variance.
2. When the experimental units are homogeneous, the design used in RBD.
3. The basis of sampling theory is randomization.
4. A functional variate for establishing a parameter is known as an estimator.
5. Degree of freedom is related to the number of independent observations.
6. Standard deviation of sample mean is given by n.
7. The smallest unit of content analysis is words.
8. A tentative explanation of an event or observation is known as hypothesis.
9. The summary and conclusion chapter in a research thesis is otherwise known as epilogue.
10. The collection of data in research falls under operational design.
11. Cone of experience was devised by Edgar Dale.
12. Concept of Law of large number' was given by Jacob Bernoulli.
13. Causal Analysis Framework was developed by Newcomb.
14. The concept of Thematic Appreciation Test (TAT) was given by R.A. Fisher.
15. The Rasch model is a psychometric model used for analysing categorical data.

ANSWERS

Multiple Choice Questions

1.(a)	2.(c)	3.(a)	4.(a)	5.(d)	6.(a)	7.(d)
8.(b)	9.(a)	10.(b)	11.(b)	12.(c)	13.(a)	14.(a)
15.(a)	16.(a)	17.(a)	18.(c)	19.(c)	20.(c)	21.(c)
22.(a)	23.(b)	24.(b)	25.(b)	26.(d)	27.(b)	28.(c)
29.(b)	30.(c)	31.(d)	32.(b)	33.(b)	34.(a)	35.(c)
36.(a)	37.(d)	38.(c)	39.(b)	40.(b)	41.(d)	42.(a)
43.(c)	44.(b)	45.(c)	46.(a)	47.(c)	48.(d)	49.(b)
50.(c)	51.(d)	52.(a)	53.(c)	54.(b)	55.(c)	56.(c)
57.(d)	58.(d)	59.(b)	60.(b)	61.(d)	62.(b)	63.(c)
64.(b)	65.(b)	66.(c)	67.(d)	68.(c)	69.(b)	70.(d)
71.(b)	72.(c)	73.(c)	74.(a)	75.(a)	76.(c)	77.(c)
78.(b)	79.(c)	80.(d)	81.(b)	82.(d)	83.(a)	84.(c)
85.(c)	86.(a)	87.(a)	88.(a)	89.(a)	90.(a)	91.(c)
92.(d)	93.(d)	94.(a)	95.(b)	96.(a)	97.(b)	98.(c)
99.(c)	100.(c)	101.(b)	102.(c)	103.(d)	104.(a)	105.(d)
106.(d)	107.(d)	108.(d)	109.(c)	110.(d)	111.(c)	112.(b)
113.(c)	114.(b)	115.(d)	116.(c)	117.(d)	118.(c)	119.(b)
120.(b)	121.(d)	122.(a)	123.(a)	124.(b)	125.(c)	126.(b)
127.(d)	128.(b)	129.(a)	130.(a)	131.(c)	132.(b)	133.(b)
134.(b)	135.(a)	136.(b)	137.(c)	138.(c)	139.(c)	140.(c)
141.(a)	142.(b)	143.(c)	144.(c)	145.(c)	146.(b)	147.(b)
148.(b)	149.(c)	150.(a)	151.(a)	152.(a)	153.(b)	154.(d)
155.(b)	156.(b)	157.(b)	158.(b)	159.(a)	160.(b)	161.(b)
162.(d)	163.(a)	164.(d)	165.(b)	166.(d)	167.(b)	168.(a)
169.(b)	170.(d)	171.(a)	172.(b)	173.(b)	174.(d)	175.(a)
176.(d)	177.(c)	178.(b)	179.(b)	180.(d)	181.(a)	182.(d)
183.(a)	184.(d)	185(a)	186.(c)	187.(b)	188.(b)	189.(a)
190.(c)	191.(a)	192.(d)	193.(b)	194.(d)	195.(d)	196.(a)
197.(b)	198.(a)	199.(b)	200.(a)			

Fill in the Blanks

1. Chapin and Greenwood
2. William Stephenson
3. W.G. Cochran
4. P. V. Young
5. Osgood
6. Reliability
7. Foam
8. Classification
9. Class interval
10. Categorization
11. Experimental design
12. Monograph
13. New Delhi
14. Research Hypothesis
15. J. L. Moreno

Mark True or False

1. True
2. False
3. True
4. True
5. True
6. False
7. True
8. True
9. True
10. False
11. True
12. False
13. True
14. True
15. True

CHAPTER 6

ADULT EDUCATION

Multiple Choice Questions

1. **Education is shaped by**

   (*a*) Society (*b*) Teacher

   (*c*) Student (*d*) Both (*a*) and (*c*)

2. **Adult learning is**

   (*a*) Problem centered (*b*) Subject centered

   (*c*) Instructor centered (*d*) Both (*b*) and (*c*)

3. **Elementary education covers children in the age group of**

   (*a*) 6 -11 years (*b*) 6-14 years

   (*c*) 10-14 years (*d*) 18-21 years

4. **The science that leads with adult learning is known as**

   (*a*) Androgogy (*b*) Pedagogy

   (*c*) Pedology (*d*) None of these

5. **To get the intelligence quotient (I.Q.)of the individual,the formula will be**

   (*a*) $\text{I.Q.} = \frac{\text{Chronological age}}{\textit{Mentalage}} \times 100$ (*b*) $\text{I.Q.} = \frac{\text{Mental age}}{\text{Chronological age}} \times 100$

   (*c*) $\text{I.Q.} = \frac{\text{Chronological age}}{\text{Mental age}} \times 50$ (*d*) $\text{I.Q.} = \frac{\text{Mental age}}{\text{Chronological age}} \times 10$

6. **The word androgogy have its root in**
   (*a*) Latin (*b*) French
   (*c*) Sanskrit (*d*) Greek

7. **"Here and Now" situation is the characteristic feature of**
   (*a*) Pedagogy (*b*) Androgogy
   (*c*) Edagogy (*d*) None of the above

8. **The term 'Andragogy' was coined by**
   (*a*) Van den Ban (*b*) P Freire
   (*c*) Alexander Knapp (*d*) D. Esminger

9. **A person is called an adult when he attains an age of**
   (*a*) 28 (*b*) 21
   (*c*) 18 (*d*) None of these

10. **The term pedagogy has its root in**
    (*a*) Persian (*b*) French
    (*c*) Latin (*d*) Greek

11. **The elements of teaching learning process are**
    (*a*) Teaching subject matter and learner
    (*b*) Teaching equipments and physical facilities
    (*c*) Both (*a*) and (*b*)
    (*d*) None of these

12. **"Education is the manifestation of perfection already in man" was written by**
    (*a*) R.N. Tagore (*b*) Swami Vivekanand
    (*c*) J. L. Nehru (*d*) Vinoba Bhave

13. **'Desire' can be enhanced with the help of**
    (*a*) A & V aids (*b*) Result demonstration
    (*c*) Method demonstration (*d*) All of these

14. **The essence of an effective classroom environment is**
    (*a*) A variety of teaching aids
    (*b*) Strict discipline
    (*c*) Lively student-teacher interaction
    (*d*) All of these

15. On the first day of his class,if a teacher is asked by the students to introduce himself,he should

(*a*) Ask them to meet after the class

(*b*) Ignore the demand and start teaching

(*c*) Tell them about himself in brief

(*d*) All of these

16. Moral values can be effectively inculcated among the students when the teacher

(*a*) Frequently talks about values (*b*) Tell stories of great persons

(*c*) Himself practices them (*d*) Both (*a*) and (*b*)

17. The essential qualities of a researcher are

(*a*) Spirit of free enquiry

(*b*) Systematization of theorizing of knowledge

(*c*) Reliance on observation and evidence

(*d*) All of these

18. Indian Institute and Advanced study is situated at

(*a*) New Delhi (*b*) Odisha

(*c*) Shimla (*d*) Chandigarh

19. 'Convention' can be achieved with the help of

(*a*) Individual contact (*b*) Group discussion

(*c*) Group Contact (*d*) All of these

20. Interest can be created with the help of

(*a*) Tour (*b*) Film

(*c*) Group discussion (*d*) All of these

21. Which of the following is the group contact teaching method ?

(*a*) Method demonstration (*b*) Bulletins

(*c*) Farm and Home visit (*d*) Result demonstration

22. Which one of the following devices used to create system where communication takes place between the instructor and learner ?

(*a*) Learning situation (*b*) Learning theories

(*c*) Teaching aids (*d*) Teaching methods

23. NCLP stands for

(a) National Council for Literacy Project

(b) National Commission for Labour Project

(c) National Child Labour Project

(d) National Child Labour Programme

24. The NLM was launched in the year

(a) 1980 (b) 1984

(c) 1977 (d) 1988

25. The first TLC was launched at

(a) Tiruchy (b) Tirupathi

(c) Kadapa (d) Kottayam

26. Human Rights Day is observed on

(a) 10th December (b) 2nd October

(c) 5th June (d) 31st May

27. MIS refers to

(a) Programme Content (b) Performance Appraisal

(c) Programme Efficiency (d) Programme shape

28. IPCL refers to

(a) Improved Pace and Cognitive Learning

(b) Improved Pace and Content of Learning

(c) Improved Pace and Curricular and Learning

(d) None of these

29. Rural welfare movement on a mass scale was

(a) Gurgaon experiment

(b) Indian village service

(c) Work at Shantiniketan

(d) None of these

30. The work at Shantiniketan with the objectives of

(a) To study rural problem

(b) To improve rural sanitation

(c) To aware interest in people for rural improvement

(d) All of these

31. Shri Rabindra Nath Tagore established a rural reconstruction institute in Shantiniketan in :

(*a*) 1925 (*b*) 1944

(*c*) 1921 (*d*) 1927

32. Rural literacy in India is showing?

(*a*) Fluctuating trend (*b*) Increasing trend

(*c*) Decreasing trend (*d*) None of these

33. In day-to-day conversation all of us use which of the following voice?

(*a*) Inert voice (*b*) Indicative voice

(*b*) Passive voice (*d*) Active voice

34. Elementary education covers children in the age group

(*a*) 18-21 years (*b*) 11-14 years

(*c*) 11-14 years (*d*) 6-11 years

35. Who is referred to as the father of "adult education " in USA?

(*a*) E.M. Roger (*b*) C.P. Davis

(*c*) E.C. Lindmen (*d*) M. Horton

36. A separate division of Adult education was set within UNESCO in the year

(*a*) 1950 (*b*) 1947

(*c*) 1958 (*d*) 1940

37. Experimental world literacy programme was launched by UNESCO in the year

(*a*) 1978 (*b*) 1972

(*c*) 1964 (*d*) 1950

38. "Nehru Literacy Award " and Tagore Literacy Award" are given by

(*a*) Indian Council of Agricultural Research

(*b*) International Adult Education Association

(*c*) Indian Adult Education Association

(*d*) Council of Scientific and Industrial Research

39. "Education is the manifestation of perfection already in man" are the words of

(*a*) Swami Vivekananda (*b*) Kuldeep Nair

(*c*) Leagans (*d*) O.P. Dahama

40. For maintaining an effective discipline in the class, the teacher should

(a) Allow students to do what they like

(b) Deal with the students strictly

(c) Deal with them politely and firmly

(d) Give the students some problem to solve

41. An effective teaching aid is one which

(a) Is colourful and good looking (b) Is visible to all students

(c) Activates all faculties (d) None of these

42. Those teachers are popular among students who

(a) Take classes on extra tuition fee (b) Award good grades

(c) Help them to solve their problem (d) All of these

43. Jan Shikshan Sansthan is a scheme sponsored by

(a) World Bank (b) Local Self Government

(c) Government of India (d) None of these

44. The scheme of Continuing Education was launched in the year

(a) 1990 (b) 1975

(c) 1995 (d) 1998

45. Jana Shikshan Nilayams were established for promoting

(a) Formal Education (b) Continuing Education

(c) Distance Education (d) All of these

46. PERCs were established in Indian Universities with the support of

(a) UNESCO (b) UNFPA

(c) WHO (d) World Bank

47. Population Dynamics refers mainly on

(a) Quality of life (b) Trends of population

(c) Poverty (d) Both (a) and (c)

48. Indicate the number of Regional Offices of National Council of Teacher Education

(a) 03 (b) 08

(c) 06 (d) 04

49. Sarva Shiksha Abhiyan was launched during the year

(a) 2001 (b) 2009

(c) 2005 (d) 2012

50. The target group of Non-Formal Education is

(a) 10-14 Age group (b) 7-14 Age group

(c) 6-14 Age group (d) 15-18 Age group

FILL IN THE BLANKS

1. Intelligence Quotient (I.Q.) was given by ______
2. Elementary education covers children in the age group_______
3. The term 'Andrology' was coined by ________
4. The science that deals with adult learning is _________
5. World literacy day was observed on ____________
6. National women literacy day is observed on ________
7. Father of learning is ________
8. The book Participatory Rural Appraisal Methodology and Applications is written by ________
9. Who first to note the phenomenon of adult education_______
10. "Education is the manifestation of perfection already in man" is said by __________
11. The concept of extension education process was given by____
12. Subject and target group specificity are characteristic feature of _______
13. The basic unit of extension work is _________
14. Learning is a internal process mainly controlled by _________
15. The last stage in the extension education process is _______

MARK TRUE OR FALSE

1. The word androgogy have its root in Greek.
2. Education is a process of interaction.
3. Adult learning is problem centered.
4. The central problem in adult learning is motivation.
5. The author of the book "Pedagogy of the oppressed" is Rolf Linton.

6. Adult literacy programme can be considered as non-formal education.
7. Effective teaching is for transforming people.
8. Bond theory of learning was advocated by C. L. Hull.
9. The learning curve in teaching follows a S-shape.
10. Education is shaped by society.
11. National Adult Education Programme (NAEP) was launched in India in the year 1978.
12. E. C. Lindmen is the father of Diffusion Simulation Research
13. "Nehru Literacy Award" and "Tagore Literacy Award" are given by Indian Adult Education Association.
14. Elementary Education covers children in the age group of 11-18 year.
15. Experimental world literacy programme was launched by UNESCO in 1964

ANSWERS

Multiple Choice Questions

1.(a)	2.(a)	3.(a)	4.(a)	5.(b)	6.(d)	7.(b)
8.(c)	9.(c)	10.(d)	11.(c)	12.(b)	13.(d)	14.(c)
15.(c)	16.(c)	17.(d)	18.(c)	19.(d)	20.(d)	21.(a)
22.(d)	23.(c)	24.(d)	25.(d)	26.(a)	27.(c)	28.(b)
29.(a)	30.(d)	31.(c)	32.(b)	33.(d)	34.(d)	35.(c)
36.(b)	37.(c)	38.(c)	39.(a)	40.(d)	41.(c)	42.(c)
43.(c)	44.(c)	45.(b)	46.(b)	47.(b)	48.(d)	49.(a)
50.(c)						

Fill in the Blanks

1. William Stung
2. 6-11 years
3. Alexander Kapp
4. Androgogy
5. September 8
6. October 2
7. Thorndike
8. Neela Mukharjee
9. Thomal Pale
10. Swami Vivekananda
11. Paul Leagans
12. Formal Education
13. Family
14. Learner
15. Reconsideration

Mark True or False

1. True
2. False
3. True
4. True
5. False
6. True
7. True
8. False
9. True
10. True
11. True
12. False
13. True
14. False
15. True

CHAPTER 7

RURAL SOCIOLOGY AND EDUCATIONAL PSYCHOLOGY

Multiple Choice Questions

1. **Authophobia is a fear of**

   (a) Fear of being alone *(b)* Fear of water

   (c) Fear of animals *(d)* fear of heights

2. **Sensation and knowledge is known as**

   (a) Remembering *(b)* Perception

   (c) Recognition *(d)* None of these

3. **In the following which is the strongest need?**

   (a) Thrust *(b)* Sex

   (c) Hunger *(d)* None of these

4. **Which is the cause of aggressive and destructive urge?**

   (a) Psyche *(b)* Libido

   (c) Death instinct *(d)* Eros

5. **Which is the incentive in youth club?**

   (a) Aspiration *(b)* Motivation

   (c) Recognition *(d)* Both *(a)* and *(b)*

6. **The book entitled "General psychology" is written by**

   (a) S. V. Supe *(b)* J. B. Gulford

   (c) J. B.Chitamber *(d)* A. R.Desai

7. **Who founded Young Men's Christian Association (YMCA)?**
   (a) Lord Beveridge (b) Fleming John
   (c) Paul Ehrlich (d) Sir George Williams

8. **The method of psychoanalysis has been propounded by**
   (a) Jane Austin (b) Sigmund Freud
   (c) William Hunt (d) None of these

9. **"Vanmahoutsva" is a campaign or movement for planting of tress. It was started in 1950 by**
   (a) C.P. Thalser (b) D.P. Thakur
   (c) E.M. Munshi (d) None of these

10. **The word" democracy" is derived from**
    (a) Latin roots (b) Greek roots
    (c) German roots (d) None of these

11. **Agrarian society is associated with**
    (a) Mechanization of agriculture (b) Cottage industries system
    (c) Primitive society (d) Cultivation of agriculture

12. **The family in an agrarian society is**
    (a) Patriarchal (b) Matriarchal
    (c) Matrilocal (d) Nuclear

13. **Role conflict is a major phenomenon of**
    (a) Primitive society (b) Agrarian society
    (c) Industrial society (d) Tribal society

14. **The term "Institution" in sociology is used to refer to**
    (a) Usages and roles which govern the human relations
    (b) The group of people living in a common territory
    (c) The system of social relationship
    (d) The symbols used in the society

15. **Groups are classified into primary and secondary groups by**
    (a) Summer (b) Cooley
    (c) MacIver (d) Sir Ronald Hill

16. **The basic criterion of social class is**
    (a) Status (b) Occupation
    (c) Residence (d) None of the above

17. The ideal types of social norms are

(a) Traditional (b) Modern

(c) None of the above (d) Both (a) and (b)

18. The degree to which the interacting individuals are similar in certain attributes is

(a) Homogeneity (b) Homophily

(c) Homophonous (d) Homophagus

19. The book "Introductory rural sociology" is written by

(a) J. B. Chitamber (b) A. R. Desai

(c) S. V. Supe (d) AA. Reddy

20. The book "Rural Sociology" is written by

(a) A.R. Desai (b) J. B. Chitamber

(c) S.C. Dubey (d) W. F. Ogburn

21. Sociology is the study of

(a) Social – political institutions (b) Political system

(c) Human Behaviour (d) Society

22. "Where there is life, there is society" are the words of

(a) August Comte (b) Max Weber

(c) MacIver and Page (d) Aristotle

23. "Sociology attempts the interpretative understanding of social man" who said

(a) MacIver (b) Marton

(c) Max Weber (d) Ogburn

24. Exogamy is

(a) Marriage within the group (b) Marriage outside the group

(c) An experimental marriage (d) Companionate marriage

25. Marriage within the caste is called

(a) Exogamy (b) Endogamy

(c) Hypergamy (d) Sagotra

26. The word "Sociology" originated from the language

(a) Latin (b) French

(c) German (d) Greek

27. A special investigation or experimentation aimed at discovery and interpretation of facts is

(*a*) Research (*b*) Practical
(*c*) Survey (*d*) Search

28. Method of studying and diagramming social relationships and making socio gram is called

(*a*) Socio method (*b*) Sociometry
(*c*) Socio gram (*d*) None of the above

29. The term "social physiology" was given by

(*a*) August Comte (*b*) Alfred Web
(*c*) Saint Simon (*d*) Max Weber

30. Filio centric family is one in which the important role is played by

(*a*) Mother (*b*) Father
(*c*) Children (*d*) Maternal Uncle

31. Caste is a class which is gained by

(*a*) Birth (*b*) Social relationship
(*c*) Power (*d*) Status

32. Socialization is a process which involving

(*a*) Training to adopt to society (*b*) Setting up the social norms
(*c*) Declaring every thing to society (*d*) None of these

33. Who expounded the bourgeois type of society ?

(*a*) Lewin (*b*) Mc Gregor
(*c*) Henry Fayol (*d*) Karl Marx

34. Which is the essential function of the family?

(*a*) Socialization of children (*b*) Satisfaction of sex needs
(*c*) Transmission of culture (*d*) None of these

35. The socio-gram is highly useful in representing a persons

(*a*) Informal relations with a structure
(*b*) Opinion about public issues
(*c*) Objective class position
(*d*) None of these

36. The 'bourgeoise' type of society was expounded by

(*a*) Max Weber (*b*) Herbert Spencer

(*c*) Karl Marx (*d*) T. H. Green

37. Sociometry was developed by

(*a*) Schramm (*b*) Moreno

(*c*) Rogers (*d*) Emile Durkheim

38. Family is included under

(*a*) Pertaining group (*b*) Primary group

(*c*) Secondary group (*d*) Tertiary group

39. Caste is a class gained by

(*a*) Status (*b*) Power

(*c*) Birth (*d*) None of these

40. Marriage within the caste is known as

(*a*) Exogamy (*b*) Endogamy

(*c*) Hypergamy (*d*) None of these

41. "Educational psychology" is written by

(*a*) L. R. Shukla (*b*) J. P. Guilford

(*c*) Skinner and Charles (*d*) S. N. Sriavastva

42. One of the basic elements in the leadership relationship is

(*a*) Faith (*b*) Task

(*c*) Motivation (*d*) None of the above

43. Person who actually initiates action in the community is

(*a*) Popular leader (*b*) Assumed leader

(*c*) Professional leader (*d*) Operational leader

44. The book "General Psychology" is written by

(*a*) S. V. Supe (*b*) J. B. Chitamber

(*c*) A. R. Desai (*d*) J. B. Gulford

45. An example of a professional leader is

(*a*) VEW (*b*) MLA

(*c*) MP (*d*) None of the above

46. **The term "Polygynandry" was coined by**
   (a) D. N. Majumandar (b) T. N. Madan
   (c) M. N. Srinivas (d) Plato

47. **The term "Social norm" was given by**
   (a) Weber (b) Sheriff
   (c) Leyman (d) Comte

48. **The term philosophy has its origin from**
   (a) Latin (b) German
   (c) Greek (d) None of these

49. **Sociology has been called "Social Physics" by**
   (a) August Comte (b) Walfred Pareto
   (c) MacIver (d) Herbert Spencer

50. **"Man is a social animal" are the words of**
   (a) Cristo (b) Aristotle
   (c) Spencer (d) Weber

51. **Who said "Individuality arises only if the community recedes"**
   (a) Mc Gregor (b) Emile Durkheim
   (c) Curt Lewin (d) Herbert Spencer

52. **Why caste system is harmful ?**
   (a) Because it prohibits inter-caste marriage
   (b) Because it separates social and political life
   (c) Because it denies social mobility
   (d) None of these

53. **The school of thought in psychology is also known as**
   (a) School of Psychology (b) Room of Psychology
   (c) Department of Psychology (d) University of Psychology

54. **Air, food, water, sleep, sex, *etc.* are cover under the**
   (a) Growth needs (b) Psychological needs
   (c) Physiological need (d) None of these

55. **The hereditary aspect of a person's emotional nature is depicted through**
   (a) Character (b) Traits
   (c) Temperament (d) None of these

56. The process of borrowing cultural items from another culture is called

(a) Socialization (b) Acculturation
(c) Unculturation (d) None of these

57. Norms without moral tones are known as

(a) Fads (b) Mores
(c) Folkways (d) Myths

58. Colonists forcing natives to become slaves is an example of social interaction is called

(a) Conflict (b) Exchange
(c) Coercion (d) Competition

59. The concept of primary groups was given by

(a) Durkheim (b) Cooley
(c) Ginsberg (d) None of these

60. The practice of a man marrying more then one women is known as

(a) Polyandry (b) Hypergamy
(c) Polygyny (d) None of these

61. Sensation and Knowledge is

(a) Perception (b) Recognition
(c) Recall (d) Attention

62. The book "I m OK You're OK" is written by

(a) Thomas Pele (b) Eric Berne
(c) Thomas and Amy Harris (d) Karl Marx

63. The theory of Laissez-faire given by

(a) Marshall (b) Leagons
(c) Adam smith (d) Bakunin

64. The term Freud is associated with

(a) Sociology (b) Psychology
(c) Rural development (d) Extension work

65. Which is the first in motivating the farmers?

(a) Farm Literature (b) Food
(c) Money (d) All the above

66. Primary groups are important for the individual because

(*a*) They help him in the spontaneous manifestation of his personality

(*b*) They provide in him a sense of unity

(*c*) They make him a social animal

(*d*) They secure him better conditions of living

67. The word "Family has been taken from the word "*famulus*" which is a

(*a*) Greek word (*b*) Roman word

(*c*) Latin word (*d*) English word

68. Family is a

(*a*) Primary group (*b*) Secondary group

(*c*) Tertiary group (*d*) All of these

69. Family is a/an

(*a*) Economic group (*b*) Religious group

(*c*) Kinship group (*d*) All the above

70. Which of the following are the informal types of sanctions?

(*a*) Ridicule (*b*) Imprisonment

(*c*) Fine (*d*) Torture

71. Who is called the farther of modern sociology?

(*a*) A. Comte (*b*) Spencer

(*c*) R. K. Meston (*d*) Max Weber

72. "Rural Sociology " is published from

(*a*) German Sociological Association

(*b*) British Sociological Association

(*c*) American Sociological Association

(*d*) None of these

73. The concept of primary and secondary groups has been given by

(*a*) Pereson (*b*) Jefferman

(*c*) Weber (*d*) C.H. Cooly

74. The term "vulgarization" is used in context of extension in

(*a*) Dutch (*b*) French

(*c*) German (*d*) India

75. The concept of psychology is known as study of soul is

(a) Modern concept of psychology
(b) Scientific concept of psychology
(c) Historical concept of psychology
(d) None of these

76. The process through which information coming from senses is "transformed, reduced, elaborated, and used" is

(a) Recognition
(b) Attention
(c) Memory
(d) Cognition

77. A 10 year old child with a mental age of 12 has an IQ of

(a) 80
(b) 42
(c) 18
(d) 120

78. IQ scores are not very dependent until the age of

(a) Nine year
(b) Eight year
(c) Seven year
(d) Six year

79. Curiosity is an example for

(a) Stimulus motive
(b) Primary motive
(c) Secondary motive
(d) Tertiary motive

80. Air, food, water, sleep, sex *etc.* cover the

(a) Psychological needs
(b) Physiological needs
(c) Growth needs
(d) None of the above

81. What is the aim of Bhoodan movement?

(a) To give land to the tillers
(b) To provide accommodation to the miserable
(c) To eliminate the possibility of the bloody revolution
(d) To orient the social reconstruction

82. Social system refers to change in

(a) Value system
(b) Dress
(c) Food habits
(d) Habitation pattern

83. Fact is "an empirically verifiable observation" who said this?

(a) Goode and Hatt
(b) MacIver
(c) Abel
(d) G.E. Swanson

84. The book " what is property is written by

(a) Karl Marx
(b) Godwin
(c) Proudhon
(d) Sanders

85. Theory proposed by Karl Marx was

(a) Anarchism
(b) Fascism
(c) Communism
(d) Socialism

86. The war between USA and Taliban is an example of

(a) Competition
(b) Conflict
(c) Assimilation
(d) Contravention

87. Caste is class gained by

(a) Power
(b) Status
(c) Birth
(d) Occupation

88. The book "Introductory Rural Sociology is written by

(a) W.F. Ogburn
(b) J.B. Chitamber
(c) A.R. Desai
(d) S.C. Dubey

89. The term sociology was coined by

(a) August Comte
(b) Karl marx
(c) Marx Weber
(d) Emile Durkheim

90. The term " society " is derived from

(a) Latin
(b) Arabic
(c) Germen
(d) Greek

91. In the following which is not classified as a psychosomatic disorder?

(a) Dermatitis
(b) Liver cirrhosis
(c) Thymus
(d) Psoriasis

92. During time of sleep arousal is

(a) Low
(b) Medium
(c) High
(d) Uncertain

93. A person who actually initiates the action in the community is known as

(a) Popular leader
(b) Operational leader
(c) Professional leader
(d) None of these

94. Which one of the following is the symbolic trait of an economic institution?

(a) Will
(b) Charter
(c) Flag
(d) Emblem

95. Element of behaviour are

(*a*) Knowledge, attitude and aspiration

(*b*) Knowledge, attitude and skill

(*c*) Knowledge, awareness and skill

(*d*) Knowledge, motivation and skill

96. The book "Introductory Rural Sociology " is written by

(*a*) S. C. Dubey (*b*) W. F. Ogburn

(*c*) J. B. Chitamber (*d*) A. R. Desai

97. Any change in the behaviour that takes place as a result of experience may be called

(*a*) Teaching (*b*) Education

(*c*) Extension (*d*) **Learning**

98. The book "Rural Sociology " is written by

(*a*) S. C. Dubey (*b*) W. F. Ogburn

(*c*) J. B. Chitamber (*d*) A. R. Desai

99. Without social contact, education is

(*a*) Not possible (*b*) Sometimes possible

(*c*) Always possible (*d*) None of these

100. In the following which is most closely related happiness ?

(*a*) Religion (*b*) Wealth

(*c*) Social relationship (*d*) Age

101. Education is a process of

(*a*) Socialization (*b*) Learning

(*c*) Institutionalization (*d*) None of these

102. A body of general principles or laws of a field of knowledge

(*a*) Principle (*b*) Mandate

(*c*) Philosophy (*d*) None of these

103. The term Philosophy has originated from

(*a*) French (*b*) Hindi

(*c*) Latin (*d*) Greek

104. Father of Sociology is

(*a*) Hegel (*b*) Auguste Comte

(*c*) Max Weber (*d*) None of these

105. Basic unit of society

(a) Neighbourhood (b) Family

(c) Group (d) Community

106. The word family have its root in

(a) Greek (b) Latin

(c) Roman (d) Sanskrit

107. The rule that one must marry within one's caste is called

(a) Exogamy (b) Endogamy

(c) Hyper gamy (d) None of these

108. The Indian type of family is characterized by

(a) Polyandrous (b) Conjugal

(c) Consanguineous (d) Matrilineal

109. Marriage due to mutual affection is known as

(a) Rakshasa (b) Gandharva

(c) Daiva (d) None of these

110. The form of marriage under which girl is given in marriage to a bridegroom with clots and ornaments is known as

(a) Paisacha (b) Asura

(c) Brahma (d) None of these

111. Sociogram is best described as a sociological

(a) Technique (b) Concept

(c) Method (d) Principle

112. A person who actually initiates the action in the community is

(a) Operational leader (b) Professional leader

(c) Assured leader (d) None of these

113. Psychology was defined as

(a) Science of mind (b) Science of soul

(c) Science of consciousness (d) None of these

114. Psychology studies the behaviour of

(a) Human and Animal (b) Human

(c) Animal (d) None of the above

115 The German word" Gestalt" means

(*a*) Elements (*b*) Shape

(*c*) Combination (*d*) **Configuration**

116. The rules for word order in sentences is known as

(*a*) Phonemes (*b*) Grammar

(*c*) Punctuation (*d*) Syntax

117. The word Androgyny literally

(*a*) Man-man (*b*) Man-women

(*c*) Women-women (*d*) None of the above

118. Psephology is word used to describe the study of

(*a*) Morphological relation (*b*) Sexual relations

(*c*) Adult delinquency (*d*) Elections

119. Thanophobia is

(*a*) Fear of fire (*b*) Fear of blood

(*c*) Fear of death (*d*) Fear of water

120. Autophobia is

(*a*) Fear of heights (*b*) Fear of blood

(*c*) Fear of being alone (*d*) Fear of water

FILL IN THE BLANKS

1. Stanford-Binet Intelligence Test developed by ________
2. Theory of Identical Elements proposed by________
3. The Book Rural Sociology in India is written by _____
4. Sociometry was developed by_______
5. The concept of primary group was introduced by _____
6. The early school of psychology was named _______
7. The most widely used statistics in psychology is __________
8. Classical experiment on conditioning was done by _______
9. The German word '*Gestalt*' means____
10. Agricultural Fundamentalism was given by_____
11. Concept of social evaluation was first given by _______
12. The idea of sociometry was first put forward by______

13. Social vaccum theory was proposed by ______
14. Socialization is a process of converting a biological organisation into _______
15. The processing of borrowing cultural items from another culture is known as _______
16. The practice of a man marring more than one women is known as _____
17. Origin of society was due to ______
18. Society is a web of social relationship according to _______
19. The movement of people outside a society is known as______
20. The rule that one must marry within one's own caste is known as _______

MARK TRUE OR FALSE

1. Day Dreaming is a artistic thinking.
2. Phrenology is associated with measurement of memory.
3. Profound memory deficit is known as amnesia.
4. Endogamy is the marriage within caste.
5. Socio-gram is best described as a sociological theory.
6. Gerontologists is the expert on the problem of ageing.
7. The memories impression left on the brain is called as memory images.
8. Thanophobia is fear of animals.
9. The word Androgyny literally means man-women.
10. Arousal during the time of sleep is high.
11. IQ scores are not very dependent until the age of six.
12. Social change means a change in social structure.
13. Role of conflict is a major phenomenon of tribal society.
14. The term Rurabinate meaning rural and urban people was coined by Donald Tuttle.
15. Social change refers to change in food habits.
16. Psychoanalysis was founded in Australia by Sigmoid Freud.
17. The term 'amnesia' means loss of memory.
18. The period of transition from childhood to adulthood is known as young age.
19. Frustration arouses hostility.
20. Biological needs are essential because they maintain homeostasis.

ANSWERS

Multiple Choice Questions

1.(a)	2.(b)	3.(a)	4.(c)	5.(c)	6.(b)	7.(d)
8.(b)	9.(c)	10.(b)	11.(c)	12.(a)	13.(c)	14.(b)
15.(b)	16.(a)	17.(b)	18.(b)	19.(a)	20.(a)	21.(d)
22.(c)	23.(c)	24.(b)	25.(b)	26.(a)	27.(a)	28.(b)
29.(c)	30.(c)	31.(a)	32.(a)	33.(d)	34.(b)	35.(a)
36.(c)	37.(b)	38.(b)	39.(c)	40.(b)	41.(c)	42.(a)
43.(d)	44.(d)	45.(a)	46.(a)	47.(c)	48.(c)	49.(a)
50.(b)	51.(b)	52.(c)	53.(a)	54.(c)	55.(c)	56.(b)
57.(c)	58.(c)	59.(b)	60.(c)	61.(a)	62.(c)	63.(c)
64.(b)	65.(a)	66.(a)	67.(b)	68.(a)	69.(c)	70.(a)
71.(c)	72.(c)	73.(d)	74.(b)	75.(c)	76.(d)	77.(d)
78.(d)	79.(a)	80.(b)	81.(d)	82.(a)	83.(d)	84.(c)
85.(c)	86.(b)	87.(c)	88.(b)	89.(a)	90.(a)	91.(c)
92.(a)	93.(b)	94.(a)	95.(b)	96.(c)	97.(d)	98.(c)
99.(a)	100.(b)	101.(a)	102.(c)	103.(d)	104.(b)	105.(b)
106.(c)	107.(b)	108.(c)	109.(b)	110.(c)	111.(a)	112.(a)
113.(b)	114.(a)	115.(d)	116.(d)	117.(b)	118.(d)	119.(c)
120.(c)						

Fill in the Blanks

1. Lewis Terman
2. Aristotle
3. A.R. Desai
4. Moreno
5. Cooley
6. Structuralism
7. Correlation Coefficient
8. Pavlov
9. Configuration
10. Davis
11. Herbert Spencer
12. Mereno
13. Sorokin
14. Human being
15. Acculturation
16. Polygyny
17. Evolution
18. MacIver
19. Emigration
20. Endogamy

Mark True or False

1. True
2. False
3. True
4. True
5. False
6. True
7. True
8. False
9. True
10. False
11. True
12. True
13. False
14. True
15. False
16. True
17. True
18. False
19. True
20. True

CHAPTER 8

ENTREPRENEURIAL DEVELOPMENT

Multiple Choice Questions

1. **The word entrepreneurship is derived from French word *Entreprendre* means**

   (*a*) To evaluate (*b*) To extend
   (*c*) To undertake (*d*) To monitor

2. **Who used first entrepreneur term**

   (*a*) George Mead (*b*) Richard Cantilon
   (*c*) D. Benor (*d*) F.L. Bryne

3. **Richard Cantilon used first entrepreneur term to refer**

   (*a*) Economic activities (*b*) Production activities
   (*c*) Physical activities (*d*) Basic activities

4. **Theory of profit was given by**

   (*a*) A.L. Bowley (*b*) R.L. Acroff
   (*c*) F H Knight (*d*) Thomas Cochran

5. **Theory of market equilibrium was propounded by**

   (*a*) Edgar Dale (*b*) F A Von Hayek
   (*c*) Thomas Cochran (*d*) Hager

6. **The term 'entrepreneurship' derived from the word**

   (*a*) *Entreprendre* (*b*) Entrepreneur
   (*c*) Micro-enterprise (*d*) Macro enterprise

7. **Which of the following is not a characteristics of an entrepreneur**

   (*a*) Decision maker (*b*) Organizer
   (*c*) Traditional (*d*) Innovator

8. **Which of the following factor has negative relationship with the rate of entrepreneurship development**

   (*a*) Lack of capital (*b*) Lack of political turmoil
   (*c*) Lack of social insurgency (*d*) Both (*b*) and (*c*)

9. **Entrepreneurs have .... level of achievement motivation**

   (*a*) High (*b*) Medium
   (*c*) Low (*d*) None of these

10. **The entrepreneurs who readily adopt the innovations inaugurated by innovative entrepreneur are known as**

    (*a*) Fabian entrepreneur (*b*) Drone entrepreneur
    (*c*) Solo entrepreneur (*d*) Imitative entrepreneur

11. **Availability of raw material is the prime issue for**

    (*a*) Enterprise location (*b*) Enterprise lay out
    (*c*) Enterprise budgeting (*d*) Capital investment

12. **Reliability of the product is the key factor of**

    (*a*) Industrial management (*b*) Product design
    (*c*) Production design (*d*) Industrial sickness

13. **Which is the last stage in the process of recruitment?**

    (*a*) Appointment (*b*) Orientation
    (*c*) Placement (*d*) Promotion

14. **Managers tends to be promoted to the level of their incompetence is said in :**

    (*a*) Parkinson's law (*b*) Marconi Law
    (*c*) Peter principle (*d*) None of these

15. **Which the essence of promotion?**

    (*a*) Change of title
    (*b*) Change of scale
    (*c*) Change of assignment
    (*d*) Change of duties and punishment

16. The basic objective of an interview is to assess the candidate's

(*a*) Intellectual capacity (*b*) General knowledge

(*c*) Personal qualities (*d*) None of these

17. The technique that gives managers a clear idea of their path of development

(*a*) Managerial appraisal (*b*) Assignment centers

(*c*) Planned progression (*d*) Work environment

18. "If it is violated, authority is undermined discipline is in jeopardy, order disturbed and stability threatened: refers mainly to the principle of

(*a*) Hierarchy (*b*) Span of control

(*c*) Coordination (*d*) Unity of control

19. Discharging of duties according to the letter and spirit of the command signifies the principles of

(*a*) Control (*b*) Authority

(*c*) Accountability (*d*) Responsibility

20. The span of control depends on

(*a*) Personality of superior (*b*) Nature of work to be supervised

(*c*) Age of the organization (*d*) All of these

21. The principle of unity of command is complimentary to the principle of

(*a*) Span of control (*b*) Management

(*c*) Scalar chain (*d*) Supervision

22. Which type of communication is known as public relation ?

(*a*) Intrapersonal communication (*b*) Interpersonal communication

(*c*) Internal communication (*d*) External communication

23. An official communication either oral or written is known as

(*a*) Informal communication (*b*) Formal communication

(*c*) Both (*a*) and (*b*) (*d*) None of the above

24. A formal communication may be

(*a*) Mandatory (*b*) Indicative

(*c*) Both (*a*) and (*b*) (*d*) None of these

25. A mandatory communication is mostly

(*a*) Vertical (*b*) Horizontal

(*c*) Both (*a*) and (*b*) (*d*) None of these

26. **The most important for an entrepreneur to have is**
    (a) High need for achievement
    (b) High knowledge about the business
    (c) High operational skills
    (d) High financial resources

27. **Technology recommended for an entrepreneurship must be**
    (a) Appropriate (b) Innovative
    (c) Traditional (d) Costly

28. **Entrepreneur are people who take**
    (a) High risk (b) Low risk
    (c) No risk (d) Moderate risk

29. **Which are the successful entrepreneurs ?**
    (a) Risky people (b) Rational people
    (c) Power hungry (d) Empire builders

30. **Who was the father of classical economics ?**
    (a) Edger Dale (b) Adam Smith
    (c) Clueny (d) H.W. Butt

31. **Leon Walrus treated entrepreneur as .... factor of production**
    (a) Third (b) Seventh
    (c) Fourth (d) Fourth

32. **The difference between current asset and current liability is**
    (a) Gross working capital (b) Net working capital
    (c) Venture capital (d) Working place

33. **PERT is a**
    (a) Probabilistic model (b) Deterministic model
    (c) Both (a) and (b) (d) None of these

34. **CPM is a/an**
    (a) Event oriented approach (b) Activity oriented approach
    (c) Both (a) and (b) (d) None of these

35. **Grouping of buyers is known as**
    (a) Marketing mix (b) Market assimilation
    (c) Market segmentation (d) Market intelligence

36. 5 Ps are important for

(a) Marketing mix
(b) Market assimilation
(c) Market segmentation
(d) Market intelligence

37. Purpose of time study is

(a) To remove wastage of time
(b) to make workers punctual
(c) To determine fair day's work
(d) To give timely assistance

38. To identify exactly what workers do and what skills or knowledge they need to succeed in a job we go for

(a) Job analysis
(b) Item analysis
(c) Factor analysis
(d) Item analysis

39. The situation that an employee must be able to cope with if he is going to succeed in a particular job is known as

(a) Critical incidents
(b) Work behaviour
(c) Work Motivation
(d) Work place

40. A lower job satisfaction will lead to

(a) High absenteeism
(b) Low moral
(c) High employee turn over
(d) Both *(a)* and *(b)*

41. Which is the most important means of coordination?

(a) Hierarchy
(b) Consultation
(c) Planning
(d) Supervising

42. Allocation of fund for different sectors of programme

(a) Controlling
(b) Budgeting
(c) Financing
(d) Controlization

43. Brainstorming is basically a

(a) Creative technique
(b) Motivation technique
(c) Evaluation technique
(d) Multiplication technique

44. Formation of cliques in an organization ultimately creates

(a) Damaging effect
(b) Performing effect
(c) Stabilizing effect
(d) Work effect

45. Measuring and correcting of activities of subordinates to ensure that events confirm to plans

(a) Supervision
(b) Controlling
(c) Managing
(d) Responsibility

46. The principle of supervision is inherent in the principle of

(*a*) Span of control (*b*) Coordination

(*c*) Hierarchy (*d*) Unity of command

47. The technique of supervision include

(*a*) The promulgation of service standard

(*b*) Budgetary limitations upon the operations

(*c*) Reporting system on work progress

(*d*) All of the above

48. An expression of discontent or dissatisfaction with any aspect of organization is

(*a*) Stress (*b*) Conflict

(*c*) Grievance (*d*) None of these

49. Source of stress in official situation is

(*a*) Overload of work (*b*) Lack of adjustment

(*c*) Lack of autonomy (*d*) All of these

50. Organizational conflict does not occur when there exists

(*a*) Autocratic supervision (*b*) Incompatible goals

(*c*) Unlimited resources (*d*) Homogeneous group

51. The informal means of coordination is

(*a*) Planning (*b*) Leadership

(*c*) Conferences (*d*) Seminar

52. A well evolved work plan designed to achieve specific objectives within a specified period of time is known as

(*a*) A project (*b*) An opportunity

(*c*) Diffusion (*d*) Conclusion

53. Social solidarity results in a) b) c) d)

(*a*) An organised society (*b*) A disintegrated society

(*c*) Both (*a*) and (*b*) (*d*) None of these

54. Liberalization of economy

(*a*) Increases competition in the market

(*b*) Increases the opportunity for entrepreneurship development in the market

(*c*) Attracts more investment in the market

(*d*) All of these

55. Women entrepreneurs emerge more in

(*a*) Male dominated society (*b*) Liberal society
(*c*) Closed society (*d*) Female dominated society

56. Existence of strict licensing rule generally

(*a*) Enhance entrepreneurship development
(*b*) Restrict entrepreneurship development
(*c*) Increases competition in the market
(*d*) All of these

57. How can manage spotting of early opportunities ?

(*a*) by Conflict (*b*) by Cooperation
(*c*) by Competition (*d*) by Fluctuation

58. Which one of the following entrepreneurship models ?

(*a*) Berlo model (*b*) Spiral model
(*c*) Evaluation model (*d*) Uniqueness model

59. Generally sustainable, community oriented and directly marketed agriculture is known as

(*a*) Agripreneurship (*b*) Intrapreneurship
(*c*) Entrepreneurship (*d*) None of these

60. Each individual in an organization must receive his or her instructions for a particular operation from only one person is known as

(*a*) Spam of control (*b*) Unity of command
(*c*) Hierarchy (*d*) Environment

61. The essence of the scalar principle is

(*a*) Unity of command (*b*) Span of control
(*c*) Communication (*d*) Coordination

62. The indicative or explanatory communication may exist

(*a*) Diagonal (*b*) Horizontal
(*c*) Vertical (*d*) All of the above

63. The communication type which provides instructions, information, and clarification known as

(*a*) Informal communication (*b*) Formal communication
(*c*) Administrative communication (*d*) Across communication

64. The communication that take place between two or more functionaries at the same level under the same supervisors

(a) Downward communication (b) Upward communication

(c) Horizontal communication (d) Grapevine communication

65. Which own of the following is not an element of communication

(a) Consistency (b) Uniformity

(c) Rigidity (d) Adequacy

66. J A Schumpter defined entrepreneur as ....

(a) Creator (b) Innovator

(c) Fabian (d) None of these

67. The National Institute for Entrepreneurship and Small Business Development (NIESBUD) is located at

(a) Noida (b) Hyderabad

(c) Bikaner (d) Lucknow

68. According to Peter Drucker entrepreneurship is a

(a) Exercise (b) Practice

(c) Theory (d) Drone

69. Danhof's classification of entrepreneur - innovating entrepreneur, adoptive/imitative entrepreneur, fabian entrepreneur and

(a) Drone entrepreneur (b) Sceptical

(c) Innovator (d) None of these

70. Fabian entrepreneur are in their approach to adopt/innovate new technology in their enterprise

(a) Innovator (b) Sceptical

(c) Fabian (d) None of these

71. Which of the following need according to Maslow comes under social need?

(a) Need for food (b) Need for friendship

(c) Need for justice (d) None of these

72. Innovating entrepreneur is one who

(a) Adopts successful innovations readily (b) Refuses to adopt new things

(c) Introduces new goods (d) None of these

73. Which are the negative factor in SWOT analysis ?

(a) S & W (b) W & O

(c) O & S (d) W & T

74. Sam Pitroda is an example of

(a) Empire builder (b) Managerial entrepreneur

(c) Mobile entrepreneur (d) None of these

75. Azim Premji is an example of

(a) Empire builder (b) Innovative entrepreneur

(c) Mobile entrepreneur (d) Managerial entrepreneur

76. J. N. Tata is an example of which are of the following ?

(a) Empire builder (b) Innovative entrepreneur

(c) Mobile entrepreneur (d) Managerial entrepreneur

77. Max Weber given the theory of

(a) Theory of Entrepreneurial Supply (b) Role Taking Theory

(c) Theory of Communication Act (d) Religious belief

78. Theory of entrepreneurial supply was given by

(a) George Mead (b) Thomas Cochran

(c) Everett E Hagen (d) J.P. Leagons

79. Theory of social change was propounded by

(a) Robert Gayar (b) J.P. Leagons

(c) Thomas Cochran (d) Everett E Hagen

80. Theory of group level pattern was propounded by

(a) F Young (b) Everett E Hagen

(c) James Stuart (d) Edgar Dale

81. .... only adopt the new tech when they realize failure to adopt will lead to loss on enterprise

(a) Sceptical (b) Outlook

(c) Innovator (d) Fabian entrepreneur

82. Fabian entrepreneur love to be remain in .... business

(a) New (b) Old

(c) Existing (d) None of these

83. .... are comfortable with their old fashioned technology of production

(a) Sceptical (b) Drone entrepreneur

(c) Outlook (d) Innovator

84. A systematic integrated and planned approach for improving enterprise effectiveness:

(a) Organizational development
(b) Futures organization
(c) Management grid
(d) Management development

85. The highest quality organization is:

(a) Transcendental
(b) Mutaualistic
(c) Parasitic
(d) Accidental

86. The relationship wherein one gains from an association at the expense of the other:

(a) Transcendental
(b) Mutualistic
(c) Parasitic
(d) None of these

87. The term "entrepreneur" is given by:

(a) Richard Cantinlon
(b) F. Young
(c) R.L. Acroft
(d) Wilsom M.

88. Who propounded Theory of X efficiency ?

(a) H.W. Butt
(b) Everett E Hagen
(c) Harvey Leibenstein
(d) J. Leagons

89. Who propounded theory of .... is also known as Gap filling theory

(a) X efficiency
(b) Y efficiency
(c) Efficiency theory
(d) None of these

90. The principles that is directly concerned with preventing overlap, conflict and friction in an organization:

(a) Direction
(b) Supervision
(c) Coordination
(d) Span of control

91. Which of the following is the limiting factor in the development of coordination:

(a) Time
(b) Size
(c) Informal organization
(d) Both *(a)* and *(b)*

92. The aspect of supervision does not include:

(a) Technical aspect
(b) Personal aspect
(c) Political aspect
(d) Theoretical aspect

93. Take the odd man out from the following

(a) Country club leadership (b) Middle road leadership
(c) Lassez-faire leadership (d) Team leadership

94. A basically lazy approach that avoids as much work as possible

(a) Impoverished management (b) Country club management
(c) Authority-compliance (d) Team management

95. The Path-Goal approach of leadership theory was formulated by:

(a) Robert House (b) Fred E Fiedler
(c) B F Skinner (d) L W Porter

FILL IN THE BLANKS

1. The term 'entrepreneurship' derived from the words____________
2. The essence of the scalar principle is____________
3. The term 'entrepreneur' is coined by_______
4. Father of classical economics was______
5. The National Institute for Entrepreneurship and Small Business Development (NIESBUD) is located at ________
6. Theory of x efficiency was given by _________
7. Theory of x efficiency is also known as _________
8. Theory of group level pattern was given by________
9. Theory of entrepreneurial supply was propounded by______
10. Generally sustainable, community oriented and directly marketed agriculture is known as_________

MARK TRUE OR FALSE

1. Richard Cantilon used first entrepreneurs term to refer economic activities.
2. The basic objective of an interview is to assess the candidate's personal qualities.
3. The last stage in the process of recruitment is promotion.
4. The principle of supervision is inherent in the principle of hierarchy.
5. the ensure of scalar principle is communication.
6. Technology recommended for an entrepreneurship must be appropriate.
7. Availability of raw material is the prime concern for enterprise location.

8. An inventor is producer or creator of ideas where as innovator implements new ideas.
9. Theory of y-efficiency is also known as Gap Filling Theory.
10. According to Peter Drucker entrepreneurship is a practice.

ANSWERS

Multiple Choice Questions

1.(c)	2.(b)	3.(a)	4.(c)	5.(b)	6.(a)	7.(c)
8.(a)	9.(a)	10.(d)	11.(a)	12.(b)	13.(b)	14.(c)
15.(c)	16.(c)	17.(c)	18.(d)	19.(c)	20.(d)	21.(c)
22.(d)	23.(b)	24.(c)	25.(c)	26.(a)	27.(a)	28.(d)
29.(d)	30.(b)	31.(c)	32.(b)	33.(a)	34.(b)	35.(c)
36.(a)	37.(c)	38.(a)	39.(a)	40.(d)	41.(c)	42.(b)
43.(a)	44.(a)	45.(b)	46.(c)	47.(d)	48.(c)	49.(d)
50.(c)	51.(b)	52.(a)	53.(a)	54.(d)	55.(b)	56.(b)
57.(c)	58.(d)	59.(a)	60.(b)	61.(a)	62.(d)	63.(c)
64.(c)	65.(c)	66.(b)	67.(a)	68.(b)	69.(a)	70.(b)
71.(b)	72.(c)	73.(d)	74.(c)	75.(b)	76.(a)	77.(d)
78.(b)	79.(d)	80.(a)	81.(d)	82.(c)	83.(b)	84.(a)
85.(a)	86.(c)	87.(a)	88.(a)	89.(a)	90.(c)	91.(d)
92.(c)	93.(c)	94.(a)	95.(a)			

Fill in the Blanks

1. Entreprendre
2. Unity of Command
3. Richard Cantlon
4. Adam Smith
5. Noida
6. Harvey Leibenstein
7. Gap Filling Theory
8. F. Young
9. Thomas Cochran
10. Agripreneurship

Mark True or False

1. True
2. True
3. False
4. True
5. False
6. True
7. True
8. True
9. False
10. True

CHAPTER 9

ELEMENTARY STATISTICS

Multiple Choice Questions

1. **A function of all sampling units in the sample is known as**

   (a) Parameter (b) Statistic

   (c) Estimator (d) Estimate

2. **The sampling fraction is defined as**

   (a) N/n (b) n/N

   (c) nCa (d) N-n/N

3. **The best sample size from a perfectly homogeneous population in respect of a character:**

   (a) A large sample (b) A small sample

   (c) A single item (d) No item

4. **Who is the author of the book 'Sampling Techniques":**

   (a) W G Cochran (b) F. A. Von Hayek

   (c) F. Young (d) Wilson M

5. **Which is the most frequent score in a given distribution**

   (a) Mean (b) Mode

   (c) Median (d) Dispersion

6. **In case of positively skewed distribution, the relation between mean, median and mode that hold is:**

   (a) Median>Mean>Mode (b) Mean>Median>Mode

   (c) Mean = Median = Mode (d) Mode > Mean > Median

7. **Which mean of dispersion is most affected by extreme values:**
   (*a*) Range (*b*) Quartile deviation
   (*c*) standard deviation (*d*) None of these

8. **The measure of dispersion which ignores signs of deviation from the central value:**
   (*a*) Range (*b*) Quartile deviation
   (*c*) Standard deviation (*d*) Mean deviation

9. **If r denotes the correlation coefficient and m denotes the slope of regression line inter-changing x-and y-axes would**
   (*a*) Change m but not r (*b*) Change r but not m
   (*c*) Change both r and m (*d*) Not change r or m

10. **The square of S.E. of an estimator is known as**
   (*a*) Sampling variance (*b*) Population variance
   (*c*) Population mean deviation (*d*) None of these

11. **Stratification gains in the**
   (*a*) Precision of an estimate
   (*b*) Sampling variance of an estimate
   (*c*) S.E. of an estimate
   (*d*) Unbiasedness of an estimate

12. **Efficiency of cluster sampling as compared with SRSWOR is**
   (*a*) More (*b*) Less
   (*c*) Equivalent (*d*) None

13. **If ten treatments are there, how many blocks are needed to have a minimum error degree of freedom of 12 in RBD:**
   (*a*) 1 (*b*) 4
   (*c*) 3 (*d*) 6

14. **A family of parametric distribution with mean is equal to variance is:**
   (*a*) Binomial (*b*) Normal
   (*c*) Poisson (*d*) Negative binomial

15. **The variance of a binomial distribution with n and p as parameters is:**
   (*a*) n (*b*) n (1 –p)
   (*c*) np (*d*) np (1–p)

16. Which of the following is not a discrete distribution:

(*a*) Binomial (*b*) Poisson

(*c*) Normal (*d*) None of these

17. Distribution wherein mean is always greater than variance is:

(*a*) Binomial (*b*) Poisson

(*c*) Normal (*d*) None of these

18. Area of critical region depends upon:

(*a*) Size of Type I error (*b*) Size of Type II error

(*c*) Both (*a*) and (*b*) (*d*) None of the above

19. Accepting a hypothesis when it is false:

(*a*) Type I error (*b*) Type II error

(*c*) Both (*a*) and (*b*) (*d*) None of the above

20. The method used to identify the units or dimensions behind the measures:

(*a*) Multiple regression analysis (*b*) Discriminant analysis

(*c*) Profile analysis (*d*) Factor analysis

21. If it is known that an event a has already occurred, the probability of an event E given A is:

(*a*) Statistical probability (*b*) Conditional probability

(*c*) Mathematical probability (*d*) Asymmetric probability

22. If A is an event the probability of A given A is equal to:

(*a*) Zero (*b*) One

(*c*) Infinite (*d*) Indeterminate

23. List of sampling units is:

(*a*) Sample population (*b*) Sampling frame

(*c*) Sample element (*d*) Sampling unit

24. The nature of relationship between sample size and sampling error is:

(*a*) Positive (*b*) Negative

(*c*) Zero (*d*) None of these

25. Difference between estimated value and actual value is measures as:

(*a*) None sampling error (*b*) Bias

(*c*) Sampling error (*d*) None of these

26. Rejecting hypothesis when it is true

(*a*) Type I error (*b*) Type II error

(*c*) Not committing error (*d*) Any one of the above

27. Factorial experiment may be conducted in

(*a*) C.R.D. (*b*) R.C.B.D.

(*c*) L.S.D. (*d*) Any of all three

28. In case N ≠ nk, to adopt circular systematic sampling the value of sampling interval k will be an integer nearest to

(*a*) n/K (*b*) N/n

(*c*) n/N (*d*)

29. If the line y on x is linear but the line does not pass through then most suitable method of estimation is

(*a*) Ratio method (*b*) Regression method

(*c*) Product method (*d*) Both (*a*) and (*c*)

30. A sampling scheme which consists of three or more stages for ultimate selection, is known as

(*a*) Multiphase sample survey (*b*) Multipurpose survey

(*c*) In complete enumeration (*d*) None of these

31. Statistical conclusions are always

(*a*) Absolutely true (*b*) Absolutely wrong

(*c*) True on average (*d*) None of these

32. Sum of measure divided by the number of the measure

(*a*) Mean (*b*) Mode

(*c*) Median (*d*) Variance

33. Greater importance to bigger items of a given series is given by

(*a*) Median (*b*) Mode

(*c*) Arithmetic mean (*d*) Geometric mean

34. Geometric mean is better than other mean when

(*a*) Data are positive as well as negative

(*b*) Data are in ratios or percentage

(*c*) Data are binary

(*d*) All of these

35. Harmonic mean is better that other means if data are for

(a) Speeds or rates (b) Heights and lengths

(c) Binary values 0 and 1 (d) None of these

36. The most important measure of dispersion used in industry for quality control

(a) Standard deviation (b) Mean deviation

(c) Coefficient of variation (d) Range

37. The best measure of dispersion among the following is

(a) Standard deviation (b) Mean deviation

(c) Quartile deviation (d) None of these

38. Which measure of dispersion is least affected by extreme values

(a) Range (b) Quartile deviation

(c) Standard deviation (d) Coefficient of variation

39. Range is not a reliable measure of

(a) Mean (b) Dispersion

(c) Median (d) Mode

40. The level of measurement associated with Mann-Whitney U test

(a) Nominal (b) Ordinal

(c) Interval (d) Radio

41. Pick the odd man out

(a) F test (b) Kvuskal Wallis test H test

(c) Friedman test (d) Z test

42. Ogives for more that type and less than type distributions intersect at

(a) Mean (b) Median

(c) Mode (d) Variance

43. Greater the distance of Lorenz curve from the line of equal distribution

(a) More is inequality (b) Less is inequality

(c) Nothing can be inferred (d) None of the above

44. If for a distribution, coefficient of kurtosis $\gamma^2 < 0$ the frequency curve is

(a) Leptokurtosis (b) Platykurtosis

(c) Mesokurtosis (d) None of the above

45. The value of coefficient of Kurtosis can be

(a) Less than 3 (b) Greater than 3

(c) Equal to 3 (d) All of these

46. A normal curve is all but not:

(a) Symmetrical (b) Continuous

(c) Unimodal (d) A symbiotic

47. If X is binomial variant with parameters n and p. if n=1, the distribution is

(a) Poisson (b) Normal

(c) Bermoulli (d) Binomial

48. The value of probability ranges from

(a) –1 to +1 (b) 0 to ∞

(c) 0 to 1 (d) None of the above

49. For any two events A and B, P(A–B) is equal to

(a) P (a)-P (b) (b) P (b)-P (a)

(c) P (b)-P (AB) (d) P (a)-P (AB)

50. Probability of two persons being born on the same day is

(a) 1/49 (b) 1/365

(c) 1/7 (d) 1/179

51. Equality of two regression coefficient can be tested with

(a) 't' test (b) χ^2 test

(c) Z test (d) ANOVA

52. A sample of 12 specimens taken from a normal population is expected to be 50 kg to test H_0: μ = 50Vs Hi: μ not = 50, the test used is

(a) Z-test (b) t-test

(c) χ^2 test (d) F-test

53. Analysis of variance utilizes

(a) F test (b) χ^2 test

(c) Z test (d) t test

54. To test the hypothesis about proportions of items in a class, we use

(a) F test (b) χ^2 test

(c) Z test (d) t test

55. The degrees of freedom for statistics for paired t test of n pairs of observations is

(a) 2 (n–1) (b) n – 1

(c) 2n – 1 (d) 3(n –1)

56. To test the agreement between observed frequencies and expected frequencies, we use

(a) F test (b) ANOVA

(c) χ^2 (d) None of these

57. For testing the significance of difference between two means, we use

(a) F test (b) Z test

(c) χ^2 (d) t test

58. The t test tends to be Z test when sample size becomes more than

(a) 100 (b) 50

(c) 30 (d) 8

59. The parent population in analysis of variance is assumed to follow

(a) Normal distribution (b) Binomial distribution

(c) Poisson distribution (d) Geometrical distribution

60. For testing the significance of correlation coefficient, we use

(a) F test (b) ANOVA

(c) χ^2 test (d) t test

61. If A and B are two events which have no points in common, the events A and B are

(a) Mutually exclusive (b) Independent

(c) Complimentary to each other (d) None of the above

62. The probability of an impossible event is

(a) 0 (b) 1

(c) ∞ (d) None of these

63. The limiting relative frequency approach of probability is known as

(a) Statistical probability (b) Classical probability

(c) Mathematical probability (d) Asymmetric probability

64. To study average rate of change in population, we use

(a) Median (b) Mode

(c) Arithmetic mean (d) Geometric mean

65. To study the average of quantity prices, we use

(a) Median (b) Harmonic mean

(c) Arithmetic mean (d) Mode

66. Harmonic mean gives weightage to

(a) Small values (b) Positive values

(c) Large values (d) None of these

67. To study the average size shoes sold in the market, we use

(a) Mode (b) Median

(c) Arithmetic mean (d) Geometric mean

68. To study the average height of plants, we use

(a) Mode (b) Harmonic mean

(c) Arithmetic mean (d) Median

69. If all the variate values are negative, the standard deviation will be

(a) Negative (b) Positive

(c) No sign (d) None of these

70. Empirical relation between quartile deviation and standard deviation is

(a) 3QD = 2 SD (b) 4QD = 3 SD

(c) 6 QD = 5 SD (d) None of these

71. The relation between standard deviation and mean deviation is :

(a) 3 MD = 2 SD (b) 5 MD = 4 SD

(c) 6 MD = 5 SD (d) None of these

72. The average of sum of squares of deviation about the mean is called

(a) Absolute deviation (b) Variance

(c) Standard deviation (d) Mean deviation

73. Which of the following is the unit less measure of dispersion

(a) Standard deviation (b) Mean deviation

(c) Coefficient of variation (d) Quartile deviation

74. Hypothesis that the population variance has a specified value can be tested by

(a) F test (b) Z test

(c) χ^2 test (d) t test

75. For testing the independence of two attributes, we use

(a) F test (b) Z test

(c) χ^2 test (d) t test

76. Degree of freedom from χ^2 in case of 4 x 3 contingency table is

(a) 12 (b) 9

(c) 6 (d) 8

77. Degree of freedom from χ^2 always lies between

(a) 0 to 1 (b) 0 to ∞

(c) –1 to +1 (d) $-\infty$ to $+\infty$

78. Multistage sampling scheme is commonly used in

(a) Large sample survey (b) small sample survey

(c) In complete enumeration (d) Multipurpose survey

79. Randomisation plays an important role in minimization of

(a) Human bias (b) Experimental error

(c) Error variance (d) All of these

80. The ANOVA technique was introduced first time in agricultural data by

(a) Prof. P.C. Mahalanobis (b) Prof. R.A. Fisher

(c) Prof. P.V. Suchatme (d) None of these

81. Correlation coefficient is considered as that of two regression coefficients

(a) Mean deviation (b) Geometric mean

(c) Square (d) Median

82. Which technique employed to find the degree of correlation?

(a) Experimental (b) Observational

(c) Statistical (d) None of the above

83. A variance shared by two or more variables

(a) Covariance (b) Correlation

(c) Common factor variance (d) None of these

84. Correlation does not demonstrate

(a) Causation (b) Effect

(c) Reason (d) None of these

85. A relationship in correlation coefficient that does not form a straight line is

(*a*) Linear relationship
(*b*) Negative relationship
(*c*) Curvilinear relationship
(*d*) None of these

86. A test used to compare the variance of two independent samples

(*a*) F test
(*b*) χ^2 test
(*c*) t test
(*d*) Z test

87. To test the hypothesis about proportions of items in a class, we use

(*a*) F test
(*b*) χ^2 test
(*c*) Z test
(*d*) t test

88. When degree of freedom of χ^2 test is more than 100, t^1 test is approximated to

(*a*) F test
(*b*) χ^2 test
(*c*) Z test
(*d*) t test

89. Who is considered as father of statistics ?

(*a*) Ronald Fisher
(*b*) Emile Durkheim
(*c*) Francis Galton
(*d*) Carl Pearson

90. Statistics serves to provide a short hand description of large amounts of

(*a*) Scores
(*b*) Data
(*c*) Points
(*d*) Formulae

91. The non-parametric substitute for the parametric 't' test

(*a*) χ^2 test
(*b*) Kvuskal Wallis test H test
(*c*) Friedman test
(*d*) Mann-Whitney U test

92. An example for one way non-parametric analysis of variance among the following is

(*a*) F-test
(*b*) Kvaskal Wallis test
(*c*) Friedman test
(*d*) χ^2 test

93. An example for two way non-parametric analysis of variance among the following is:

(*a*) F-test
(*b*) Kvaskal Wallis test
(*c*) Friedman test
(*d*) t test

94. A functional variate for establishing a parameter is called

(*a*) An estimate (*b*) An estimator

(*c*) A frame (*d*) A statistic

95. The idea of randomization was put forwarded by

(*a*) Chapman and Greenwood (*b*) Kerlinger

(*c*) Ronald A Fisher (*d*) Sukathmae

96. When the experimental units are homogenous, the design used is

(*a*) CRD (*b*) RBD

(*c*) LSD (*d*) None of the above

97. Degree of freedom is related to the number of

(*a*) People involved (*b*) Hypothesis under test

(*c*) Independent observations (*d*) χ^2 test

98. The experimental design wherein the heterogeneity in the experimental units are in two directions

(*a*) CRD (*b*) RBD

(*c*) LSD (*d*) BIBD

99. The error d.f. for RBD experiment with 5 treatments and 3 replications is

(*a*) 15 (*b*) 18

(*c*) 8 (*d*) 10

100. It there are ten treatments and three replications in a CRD experiment, the error degrees of freedom is

(*a*) 16 (*b*) 20

(*c*) 26 (*d*) 32

101. Equality of several normal population means can be tested by

(*a*) F test (*b*) χ^2 test

(*c*) Z test (*d*) ANOVA

102. Kruskall Wallis analysis of data meant for

(*a*) No classification (*b*) One way classification

(*c*) Multi way classification (*d*) Two way classification

103. The area under standard normal curve beyond Z=+ 1.96 is

(*a*) 95 per cent (*b*) 99 per cent

(*c*) 5 per cent (*d*) 10 per cent

104. The t' test is used to test the significance of

(*a*) Two means for large sample (*b*) Two mean for small sample

(*c*) Correlation coefficient (*d*) Both (*b*) and (*c*)

105. If the measure of Kurtosis is greater than 3, then the curve is

(*a*) Platykurtic (*b*) Mesokurtic

(*c*) Leptokurtic (*d*) One of these

106. In case of a positively skewed distribution

(*a*) Median > Mean > Mode (*b*) Mean > Median > Mode

(*c*) Mean = Median = Mode (*d*) None of the above

107. The AM of the age of 20 students in a class in 15 year. When the teachers age also is added, the mean increase by one year. The age of the teachers is

(*a*) 26 Years (*b*) 24 Years

(*c*) 36 Years (*d*) 25 Years

108. Which is not a measure of central tendency

(*a*) Mean (*b*) Median

(*c*) Mode (*d*) Variance

109. Mean Deviation is minimum when taken about

(*a*) Mean (*b*) Median

(*c*) Mode (*d*) Zero

110. The relation between arithmetic mean, geometric mean and harmonic mean is

(*a*) AM=GM=HM (*b*) AM>GM>HM

(*c*) AM<GM<HM (*d*) None of these

111. Measures of dispersion includes the range and

(*a*) Standard deviation (*b*) Mode

(*c*) Median (*d*) Mean

112. Standard error can be

(*a*) 0 (*b*) 1

(*c*) > 0 (*d*) All of these

113. Unit of variance of a data in cm is

(*a*) No unit (*b*) cm^2

(*c*) cm (*d*) cm^3

114. When r = +1, the two regression lines are

(*a*) Coincident (*b*) Parallel

(*c*) Perpendicular to each other (*d*) Perpendicular to x axis

115. Usually the explanatory value of a regression equation of y on x is measured using

(*a*) Correlation coefficient (*b*) Coefficient of determination

(*c*) Coefficient of explanation (*d*) Range

116. F-test is used to test

(*a*) Ratio of two variances (*b*) Ratio of two means

(*c*) Square of normal variate (*d*) None of the above

117. Square of a standard normal variate is

(*a*) F (*b*) χ^2

(*c*) Z (*d*) t

118. Equality of several variances is done by

(*a*) F (*b*) χ^2

(*c*) Z (*d*) t

119. Standard deviation of sample mean is given by

(*a*) σ (*b*) n

(*c*) σ /root n (*d*) σ/n

120. An example for non parametric test from among the following is

(*a*) Z test (*b*) t test

(*c*) F test (*d*) χ^2 test

121. The ratio of difference between sample variance and with in sample variance follows

(*a*) Z distribution (*b*) t distribution

(*c*) F distribution (*d*) χ^2 distribution

122. In an Incomplete Block Design (I.B.D.), the number of treatment under study is

(*a*) Less than block size (*b*) Equal to block size

(*c*) Greater than block size (*d*) Equal of number of blocks

123. Factorial experiment is

(*a*) Not a design (*b*) A design

(*c*) A design and experiment both (*d*) None of these

124. Analysis of covariance is an extension of

(a) Student's test
(b) Variance ratio (F) test
(c) Analysis of variance technique
(d) Chi square (x^2) test

125. Duncan's multiple range test is more popular in

(a) Medical science
(b) Agricultural science
(c) Mathematical science
(d) Physical science

126. When for transformed data, it is not possible to make its mean and variance independent, then one should apply

(a) ANOVA technique
(b) Regression analysis
(c) Weighted analysis
(d) None of these

127. ANOVA technique is also used in testing the significance of

(a) Correlation ratio
(b) Correlation coefficient
(c) Regression coefficient
(d) Homogeneity of variance

128. In an incomplete 3 way layout experiment, the three factors responsible for variation in observation are

(a) Row, column and blocks
(b) Row, column and replication
(c) Row, column and treatment
(d) Row, treatment and blocks

129. The principle of experimental design include:

(a) Randomization
(b) Local control
(c) Both (a) and (b)
(d) None of these

130. Probability of a leap year will have 53 Sundays is:

(a) 1/7
(b) 2/7
(c) 2/53
(d) None of these

FILL IN THE BLANKS

1. In an ideal classification number of classes should be between the ranges_______
2. Characters can be classified as______
3. A quantity that varies from individual is called______
4. Number of leaves per plant is discrete variable whereas height of the plant is ___________ variable.
5. The classification of flower according to their colour is _____type of classification.

6. ____mean is a best measure of central tendency.
7. Arithmetic mean, median, mode, geometric mean, harmonic mean, weighted mean are there important measures of ________
8. Height of rectangular in a histogram is proportional to _______
9. The class having maximum frequency is known as______
10. In ________distribution values of mean, median and mode are same.
11. Standard deviation is absolute measure of depression where as_______is a relative measure of dispersion.
12. Square of standard deviation is called ______
13. Square root of variance is called__________
14. Coefficient of variation = Standard deviation ÷ ______
15. For qualitative characters we use _____method of correlation
16. for quantitative characters we use ______ method of correlation.
17. The correlation between high of the experimenter and grain yield of the crop of the experiment, which he has conducted is __________ correlation.
18. Karl Pearson's and _______ are two methods for study of correlation coefficient between two characters.
19. Square of the correlation coefficient is called_________
20. Representative part of the population is called _______
21. Mean binomial distribution is = __________
22. Standard deviation of binomial distribution = σ = ______
23. Mean and variance of Poisson distribution = γ , σ = _______
24. For normal distribution mean, median and mode are __________
25. Total area under probability curve and x axis is considered to be equal_______
26. Standard deviation of sample means is called _________
27. Null hypothesis is a statement about___________
28. To compare two population means we use _________ test.
29. To compare two population variances we use ______test.
30. To test association between two characters we use__________

MARK TRUE OR FALSE

1. For characters which can be measured by some units we use qualitative classification.
2. Arithmetic mean can be obtained by graphical method.

3. The value of arithmetic mean is not affected by fluctuation of value of extreme items.
4. Algebraic sum of squares of deviation taken from arithmetic mean is always zero.
5. Algebraic sum of squares of deviation taken from median is always minimum.
6. Standard deviation is relative measure of dispersion whereas coefficient of variation is absolute measure of dispersion.
7. Small value of standard deviation indicates that there is consistency in the values of character.
8. Value of standard error increases as sample size increases.
9. If value of correlation coefficient is 1.25 we say that there is no correlation between the two characters as the value is not satisfying the range of correlation coefficient.
10. Karl Pearson's correlation coefficient is unaffected by change of origin and scale.
11. Binomial distribution is continuous probability distribution.
12. For binomial distribution mean, median and mode is same.
13. F test is used for comparisons of two variables.
14. 't' is used for testing association between two attributes.
15. X^2 test is used to compare two variables.
16. Null hypothesis is the statement about population parameters.
17. Null hypothesis is the statement about statistics.
18. To increase the precision of conclusion we should decrease the value of level of significance.
19. A river has mean depth of water of two feet so that any adult can go through it very easily without swimming.
20. Arithmetic mean is a best measure of dispersion.
21. The median weight of fifteen students is 40 kg four move students of weight 43 kg, 39 kg 46 kg and 38 kg are added to the group. Then value of median and mean will remain same for the whole group.
22. If the sum of squares of deviation parallel to X axis is minimized then the line of best fit is called as the line of regression of X on Y.
23. If the difference sample mean and population mean is found to be significant the sample doest not belong to the population.
24. Coefficient of correlation is affected by change of origin but not by change of scale.

25. Degree of freedom of a statistic is the number of independent comparisons on which it is bases.
26. 'F' test is used for testing the significance of difference between two means while 't' test is used for testing the ratio of two variances.
27. Binomial distribution is a continuous probability distribution.
28. Coefficient of correlation indicates the degree and direction of the relationship between the variations of two variables.
29. Correlation coefficient is the geometric mean between the regression coefficient.
30. Standard deviation is a measure of variation used for comparing the variation between two series.

ANSWERS

Multiple Choice Questions

1.(b)	2.(b)	3.(c)	4.(a)	5.(b)	6.(b)	7.(a)
8.(d)	9.(a)	10.(a)	11.(a)	12.(b)	13.(c)	14.(c)
15.(d)	16.(c)	17.(a)	18.(a)	19.(b)	20.(d)	21.(b)
22.(b)	23.(b)	24.(b)	25.(c)	26.(a)	27.(d)	28.(b)
29.(b)	30.(b)	31.(c)	32.(a)	33.(c)	34.(b)	35.(a)
36.(d)	37.(d)	38.(b)	39.(b)	40.(b)	41.(d)	42.(b)
43.(a)	44.(b)	45.(d)	46.(d)	47.(c)	48.(c)	49.(d)
50.(c)	51.(d)	52.(b)	53.(a)	54.(c)	55.(b)	56.(c)
57.(d)	58.(c)	59.(a)	60.(d)	61.(d)	62.(a)	63.(a)
64.(d)	65.(b)	66.(a)	67.(a)	68.(c)	69.(b)	70.(a)
71.(b)	72.(b)	73.(c)	74.(c)	75.(c)	76.(c)	77.(b)
78.(a)	79.(a)	80.(b)	81.(b)	82.(c)	83.(c)	84.(a)
85.(c)	86.(a)	87.(c)	88.(c)	89.(a)	90.(b)	91.(d)
92.(b)	93.(c)	94.(b)	95.(c)	96.(a)	97.(c)	98.(c)
99.(c)	100.(b)	101.(a)	102.(b)	103.(c)	104.(d)	105.(c)
106.(b)	107.(c)	108.(d)	109.(b)	110.(b)	111.(a)	112.(d)
113.(b)	114.(a)	115.(b)	116.(a)	117.(b)	118.(b)	119.(c)
120.(d)	121.(c)	122.(c)	123.(a)	124.(c)	125.(b)	126.(c)
127.(a)	128.(c)	129.(c)	130.(b)			

Fill in the Blanks

1. 5 to 15
2. Quantitative and qualitative
3. Continuous variable
4. Continuous
5. Qualitative
6. Arithmetic
7. Central tendency
8. Frequencies
9. Model Class
10. Normal or symmetrical
11. Coefficient of variation
12. Variance
13. Standard deviation
14. Arithmetic mean x 100
15. Rank correlation
16. Karl Pearson's
17. Zero
18. Spear mans rank correlations
19. Coefficient of determination
20. Sample
21. np
22. $\sqrt{npq}$
23. $\sqrt{\gamma}$
24. Same
25. One
26. Standard error
27. Equality
28. 't'
29. 'f'
30. χ^2 test

Mark True or False

1. False
2. False
3. False
4. False
5. False
6. False
7. False
8. False
9. False
10. True
11. False
12. False
13. True
14. False
15. False
16. False
17. False
18. False
19. False
20. False
21. True
22. False
23. False
24. False
25. True
26. False
27. False
28. True
29. True
30. False

Glossary

Adaptability Principles in use of Extension Teaching Methods: Extension worker should have knowledge of extension methods so that they can select proper method according to the condition. Teaching methods should be flexible so that they can be properly applied on people according to then age groups, educational background, economic standard and gender.

Agricultural Extension: Agricultural extension was once known as the application of scientific research, knowledge, and technologies to improve agricultural practices through farmer education. The field of extension now encompasses a wider range of communication and learning theories and activities (organized for the benefit of rural people) by professionals from different disciplines.

Agricultural Innovation Systems (AISs): An innovation system can be defined as a network of organizations, enterprises, and individuals focused on bringing new products, processes, and forms of organization into economic use, together with the institutions and policies that affect their behavior and performance. The agricultural innovation systems concept embraces not only the suppliers of new technologies but is also concerned with the role and interaction of different actors within agricultural innovation systems, especially in connecting with new and emerging markets for different types of high-value crops and products.

Agricultural Technologies: Until recently, agricultural technologies have largely been created and disseminated by public research institutions. However, during the past 50 years, the private sector has played an increasingly important role in producing and selling proprietary technologies in the form of production inputs, such as hybrid seed, pesticides, and mechanical technologies.

Agricultural Technology Management Agencies (ATMA): The Department of Agriculture and Cooperation, Govt. of India, intends to try a new model through establishment of Agriculture Technology Management Agencies (ATMA) in selected districts involving certain stages under the National Agricultural Technology Project (NATP).

Agro-Clinic Yojna: According to this project, the latest research knowledge of agriculture will be given directly to the farmers. The unemployed agriculture graduates will organize these centers. By this clinic project, many programmes will be organized for the farmers, such as the latest agriculture technology, verified or approved seeds, the knowledge of preparing the soil for sowing different crops, the use of pesticides and insecticides, crop diseases, and their treatment, different irrigation systems, the proper use of fertilizers and water, the knowledge of agricultural instruments and the improvement of animals' variety etc. programmes.

Appropriate Technology: Appropriate technology refers to a technology package, which must be technically feasible, economically viable, socially acceptable, environment- friendly, consistent with household endowments, and relevant to the needs of farmers. Technologies are subject to adjustment, change and evolution.

Baroda Village Reconstruction Project: Shree B.T. Krishnamachari in Baroda in the Gujarat State initiated this Project in 1932. This Project Continued for a long period.

Change in Social Attitude: The rural environment in India is filled with jealousy, hatred, ill felling and conflicts. Selfishness is increasing day by day and it is harmful for the society and the country. "Extension worker should try to bring change in this mental attitude of people so that they can sacrifice their selfishness for the welfare of the society and country and may become partners in the reconstruction of the country.

Child Development Programmes: On the basis of the census 2001, the population of women and children is 49.87 crore (48.2%) and 86 lakh (15.42%) respectively out of the total population of the country. Women and Child Development Department is working under the Ministry for Human Resources and determines the programmes, planning and policies related to the development of women and children and the action plan between the government and the non-government organizations. There is also planning for women development and securing jobs for them.

Command Area Development Programme (CADP): Including three annual projects (1966-67), this type of programme was the first one made. The main object of this project was to utilize the available water in proper manner.

Commodity-Based Advisory Services: Commodity-based advisory services are similar to value-chain extension systems (defined later in this glossary), in which an economically important crop or product, generally for export (e.g., cotton, coffee, or other high-value crops or products), requires that producers use specified genetic materials or varieties and follow strict quality-control standards in producing and harvesting the crop or product.

Cooperative Extension Service: The Cooperative Extension Service is a joint effort of national, state, and county governments within the United States to advance the practical application of knowledge through a wide variety of extension and outreach activities. At the present time, this system pursues the following program areas: youth development (4-H), agricultural and rural development, natural resource management, family and consumer sciences, and community and economic development (i.e., helping local governments investigate and create viable economic options for community development).

Desert Development Programme (DDP): This programme has been initiated in 1977-78. Its main aim is integrated development of desert area. This programme is operational in 227 blocks of 36 districts of seven states i.e. Rajasthan, Haryana, Gujarat, and Jammu & Kashmir. It includes area of 4.57 lakh square km.

Drought Prone Area Programme (DPAP): In India, there is a very big drought prone area where it rains sparsely. In these areas famine generally occurs. Central government had initiated Drought Prone Area Programme for these areas in 1973-74. This programme was introduced in 947 blocks of 149 districts of 13 states.

Emphasis on Self-dependence: Our villages were self-dependent before British rule. Their needs like food, clothing and housing, were fulfilled in villages itself, but now they have to go outside to fulfill their basic needs and still they are unable to earn enough for themselves. Therefore, it is necessary to bring change in this situation and to teach lesson of self-independence to the villagers.

Employment Prone Programmes: The most burning problem of the large developing countries is the unemployment of which the other name is poverty. The basic aim of all the programmes such as industrialization, land improvement, increment in the national growth and five year plans, is that the people of the country may get jobs and their economic condition may be improved.

Etawah Pilot Project: This is such an effective project that after witnessing its results, the way was cleared for initiating the Community Development Project. Therefore this was called the Pilot Project.

Extension Education: Extension education primarily focuses on the teaching-learning methods needed to train and to provide small-scale and women farmers with the necessary skills, knowledge, and information they will need to increase their farm income and thereby improve the livelihoods of their rural families.

Extension: The word extension is mostly used for Extension Education.

Extension Educational Process: The extension process is working with the people with their immediate needs and interests, which can make available additional occupation, make improvement in the socio-economic status, better home-management and expedite welfare of the rural people.

Extension Job: The job of extension in agriculture and home science is to assist people engaged in farming and home-making to utilize their own resources more effectively and those that are available to them, in solving the current problems and in meeting the changing economic and social conditions.

Extension Service: It means an organization and/or a programme for the welfare and development, which employs the extension educational process for the implementation of programme. It is thus same as that of extension work except that in extension service there has been greater emphasis on service.

Extension Work: It means the whole structure of Extension work. It includes the process of Extension Education i.e. the process of teaching and learning. Besides the process, in extension work are included organization, administration, supervision, finances as well as the programs for the overall development.

Farmers Development Projects: For the progress of rural community, the successive governments initiated the projects. Though prosperity and development has been noticed, yet the socialists and economists attracted the government's attention towards the flourishing inequity, which had been created due to the concentration of the benefits to the medium and high-class farmers. Consequently there was no significant improvement in the conditions of small farmers and the farming labourers.

Firka Vikas Yojana: The government of Madras (now Tamil Nadu) decided to make efforts for the development of villages at Firka level. The first programme began in 1946. Among Pre-independence project, this was the biggest project.

Formal Education: It is institutionalised, chronologically graded and well-structured system of education, which starts from schooling to the higher education.

Four-H Clubs: Four-H clubs are youth organizations with the mission of "engaging youth to reach their fullest potential while advancing the field of youth development". These organizations serve over 6.5 million members in the United States, and 4-H clubs or similar organizations now exist in many other countries (see http://www.national4-hheadquarters.gov/about/4h_atlas.htm). The goal of 4-H is to develop citizenship, leadership, and life skills of youth, primarily through experiential learning programs. Though typically thought of as an agriculturally focused organization, 4-H today encourages both rural and urban members to learn about many topics, such as youth leadership, youth–adult partnership, working together to achieve common objectives, entrepreneurship, parliamentary procedures, and public speaking. The four H's stand for head, heart, hands, and health.

Full use of Present Local Resources: In order to enhance the extension work and rural industries like agriculture and cottage industries, it is necessary to efficiently utilities the available local resources so that people can become self-dependent.

Grow more Food Campaign: This campaign, started in 1942, was continued after getting the independence. The main object of this campaign was to fulfill the need of food, which had been created due to the Second World War. This campaign was the first one to be organized on a national level. In this campaign, the new seeds and chemical fertilizers were distributed among the farmers. Agricultural departments of state governments organized this campaign.

Gurgaon Project: In Gurgaon district, this programme of village development was the first one to be run by the State. It was started by the Mr. F.L. Brayne. In 1920, Mr. Brayne had been appointed on the post of Deputy Commissioner in Gurgaon district and he began this project of rural upliftment in his district, which became famous as "Gurgaon Project". The main objectives of this project were: (1) To increase crop production, (2) to control extra expenditure, (3) to improve the health, (4) to develop the feeling of women-education, and (5) home development work.

High Yielding Variety Programme (HYVP): In India, this programme has been initiated in 1963, when the Director, International Maize and Wheat Improvement Centre, Mexico Dr. N.E. Borlaug visited the fields of wheat in India and sent the 4 types of dwarf seeds of wheat for testing

Human Beings Learn Mainly by three Methods: (i) Through informal education; (ii) formal education; and (iii) non-formal education.

Ideal Village Project: In 1903, Sir Daniel Hamilton began a project of rural development on the basis of cooperation and started this project in "Sundaram Village" near Madras (Chennai). In 1910 under this programme, Co-operative Deposit Institute was established. Along with the savings, the programmes of health, literacy and small industry were started. In this planning, the emphasis was on specialized training for unemployed young farmers to make them self- dependent. This project continued till 15-16 years and after some time it disappeared.

Indian Village Service (IVS): In 1945, under the guidance of Dr. W.H. Wisher this service began in the village Agasoli, District Aligarh in U.P. But, after the partition of the country, the village volunteer M.V. Siddiqi Khan went to Pakistan and this centre was closed.

Indira Mahila Yojana: This Yojana was started on 20 August 1995 in 200 Blocks related to the Children Development Scheme. Its aim is to develop the decision taking capacity among the rural women and to make them aware of their rights. Anganwadi project has been started in the town and rural area for organizing these women. In the Anganwadi centers, Indira Mahila Kendra (IMK) has been developed where Indira Mahila Block Society is organized.

Informal Education: It is the never-ending process by which an individual learns through daily experiences and exposure to environment at home - at work, from friends, radio, television, papers and books etc.

Information and Communications Technology (ICT): Information and communications technology is an umbrella term that includes all types of technologies for the communication of information. It encompasses any medium to record and broadcast information, as well as technologies for communicating information through voice, sound, and/or images. Information technology (IT) has become a hub for communicating information, most often using computers.

Innovation System: Innovation can be defined as a new way of doing something, ranging from changes in the way we think, to the way we produce new products or use new processes or procedures. It also includes institutional innovations that change the way an organization carries out new or different functions, for example, shifting toward a bottom-up rather than a top-down extension system; or moving toward a more market-driven rather than a technology-driven extension system.

Integrated Child Development Service (ICDS): A National Policy for Children was launched in 1974. During 1975-76 under this policy ICDS Policy was launched in 33 blocks of the country and till 31st March 1996, 5,615 ICDS Projects reached to the poor areas of 310 big cities and 5291 community development blocks of the country. According to this scheme up to 6 year old children are selected. Today, this project has provided Pre-school Education to 10.5 million children (3-6 yrs). For the children of the country, this is the main programme. The services of ICDS have been provided to 22 million children and to their mothers till the year 1996 which also include providing the balanced diet to them.

Integrated Rural Development Programme (IRDP): "Integrated rural development programme" is the latest programme that has been made after considering all the success and failures of all the programmes that was introduced before and after Independence. This programme was introduced in 2,300 blocks of the country in 1978-79 and after some time, on 2nd October 1980, it was introduced in all the blocks of the country.

Intensive Agriculture Area Programme (IAAP): Encouraged by the success of Intensive Agriculture District Programme, the Centre Government introduced Intensive Agriculture Area Programme in 114 districts of country in 1964-65. Considering the distribution area of crops, those selected areas and crops were selected, which involved a greater possibility of development than others.

Intensive Agriculture Development Programme (IADP): In 1959-60 Ford Foundation's Agriculture Food Production Committee had presented some firm suggestions to the government pointing out the laxity of agriculture production programmes in the report 'India's Food Crisis and Steps to Meet It". In this report arguing on the topic of food production there was the main suggestion of community efforts by the intensive programme on the crops of food grain in selected areas.

Intensive Agriculture District Programme: According to the Intensive Agriculture Development Programme, it was emphasized that the agriculture production programme should be introduced without delay on the selected crops in selected areas. On this basis, the Central Government introduced Intensive Agriculture District Programme in 7 districts in various states of the country in 1960-61. These districts were Tanjore (Madras, now Tamil Nadu), West Godawari (Andhra Pradesh), Sambhalpur (Orissa), Raipur (M.P.), Ludhiana (Panjab), Aligarh (U.P.),and Pali (Rajasthan).

Jawahar Gram Samridhi Yojana (JGSY): The central government had initiated this project of 'Jawahar Gram Samridhi Yojana' on 1st April, 1999 in place of "Jawahar Rozgar Yojana." According to this project, it is necessary to make annual action plan and to determine its implementation through Gram Panchayat. This project includes JRY, NREP, and RLEGP etc. programmes.

Jawahar Rozgar Yojana (JRY): This project was initiated on the birth anniversary of the first Prime Minister of India, late Shri Jawaharlal Nehru during 1989-90.

Learning: The development and change in behaviour of a person which occur in relation with the attainment of a certain objectives is called learning. Meaning of learning is the development of reconciliation method for certain problem, which provides new techniques for completing the work. Learning is such a process through which consciousness is fostered, meaning is elaborated, decisions are taken and work is completed.

Mahila Samriddhi Yojana: This project is running in the entire country from 2nd October 1993 by the established post office in the 1.32 lakh rural area. In this project rural adult women are encouraged to save money. In this project, all the adult rural women deposit their income by opening their accounts in nearby post office and maximum Rs. 300 can be deposited in a year on which the Central Government provides 25% internal help after one year. 169.06 lakh accounts had been started in different rural post offices in which Rs. 172.42 crore had been deposited up to March 1996. For the popularity of this project, help of different N.G.Os have been taken.

Marginal Farmers And Agricultural Labour Development Agency (MFALA): The Government of India made efforts to bring the change in the economic condition of marginal farmers and agricultural labourers. In 1967 this project tried to provide intensive agricultural methods, loans, and production materials for the increment in their income. Inclusive this programme there was planning of the small industry for providing more chances of employment to the marginal farmers and agricultural labourers.

Marthendom Project: This project of rural development was initiated in the village Marthendom near Trivenduram of Kerala State by Young Men Christian Association (Y.M.C.A) and Christian Church Association under the direction of Dr. Spencer Hatch in 1928. This village was undeveloped economically and the economic condition of the native majority was poor.

Methods of Education: Education must be conceived as a life long learning process.

Million Well Project: In 1988-89, a new project was started for providing the facility of irrigation from the wells without any cost for such people like small and marginal farmers living under the poverty line, scheduled caste and tribal people and labourers. The name of this project was Million Well. This was introduced under the 'Employment Guarantee Programme' and the 'National Rural Employment Programme' for the landless people. In April 1989, joining the both employment programmes, a new programme was formed which was named as 'Jawahar Rozgar Yojana'. Million Well Project is being guided under 'Jawahar Rozgar Yojana'.

Minimum Needs Programme (MNP): India is a country of villages where 80% of the population resides. For the integrated development of the country we cannot overlook the development of the villages. So, after independence, emphasis was laid upon the integrated development of the villages. For this, there were introduced many programmes. Among them the main "Minimum Needs Programme" was initiated in 1977. In this programme we pay our attention to the social investment.

National Agricultural Technology Project (NATP): To make the technology development project a permanent feature under the National Agricultural Research Project and National Agriculture Extension Projects and for popularising and fulfilling the present and necessary needs of it, the Government of India took a decision to organize a new National Agricultural Technology Project with the financial support of the World Bank. This project was introduced in the financial year 1995-96. Govt. of India approved this project in November 1998 for full scale implementation.

National Agriculture Research Project (NARP): National Agriculture Research Project has been introduced for re-strengthening the Zonal Research Centre of State Agriculture University by ICAR, New Delhi, which is economically financed by the World Bank through International Development Association (I.D.A.) This project, had been introduced in 1988. The main objectives of this project were to study the agriculture related problems associated to different climates and to bring the solution after research and above all to strengthen the Zonal Research Centers.

National Demonstration (ND): Among the methods of Extension, demonstration is the most popular medium. Due to the greater impact of sense of vision, the demonstration methods are initiated in the agricultural extension. The first 'Result Demonstration' was started in Texas of America in 1903, where the farmers received the knowledge of new agricultural methods, developed instruments, seeds, and the technology related to crop protection.

National Extension Service (N.E.S.): In 1953, Grow More Food Committee had submitted its report, which suggested introducing "National Extension Service" in the whole country. This service was initiated on 2nd October 1953, so that within 10 years, the development work of agriculture could properly progress in the entire country.

National Rural Employment Programme (NREP): In 1977, for providing employment to the poor villagers and for the construction of permanent community property, programme of "Grain for the sake of Work" was started and later in 1980 it was renamed as "National Rural Employment Programme". From 1st April 1981, the central and the state governments have introduced this project on the basis of ratio of 50: 50. The wages was not given only in the form of grains but everyday per man was paid 1 kg. of grain and remaining wage was paid in cash.

Natural Resource Management (NRM): Natural resource management can be defined as the responsible and broad-based management of the land, water, forest, and biological resources base—including genes—necessary to sustain agricultural productivity and avert degradation of potential productivity.

Nilokheri Project: In 1948, Shree S.K.Dey prepared this project for the purpose of providing residence for 700 immigrants from Pakistan. He began this project using 100 acre of swampy land spreading in the midst of Karnal and Kurukshetra. The name of this project was "Majdoor Manzil". The director of this project was Shree S.K. Dey. He went on to become the Union Minister of Community Development in 1965.

Non-Formal Education: It has been a well organized, systematic educational activity that is carried on outside formal educational system in order to provide certain selected type of learning to the selected group of individuals which include adults, young as well as the children. Included in the non-formal education are subjects on the various aspects of agriculture and related fields, like training programs for farmers, farm women and rural youth.

Nongovernmental Organizations (NGOs): Nongovernmental organizations are legally constituted organizations created by private individuals or organizations with no participation or representation by any government agency. NGOs can be categorized into two types: operational and advocacy. The primary purpose of an operational NGO is to design and implement development-related projects.

Participatory Rural Appraisal (PRA): Participatory rural appraisal is a label given to a family of participatory approaches and methods that emphasize local knowledge and enable local people to make their own appraisal, analysis, and plans. The key tenets of a PRA are participation, teamwork, flexibility, and triangulation to ensure that information is valid and reliable.

Plan for all-round Development: Extension worker should organise social, economic and educational development programmes for all round development of the society.

Principle of Applied Science and Democratic Approach: Extension Education is based upon democratic principles. It is based on discussions and suggestions. Discussions are held with the people on actual condition so that they participate in work.

Principle of Cooperation and Participation: Principle of cooperation and participation is very important in extension education. Ultimately without the cooperation of people the work cannot be successful and desired result cannot be achieved. The first task of extension education is the cooperation of people and their participation in work. People should realize that the task of extension education is their own task. Participation in extension work generates confidence among people for the work.

Principle of Cultural Difference: Rural people have high faith in traditional customs and values. Therefore, extension worker should be aware of what the rural people know, Think and also about their belief. Cultural changes should be gradual and in accordance with the cultural status of people.

Principle of Encouragement: In extension work, principles of encouragement have great importance. Under pressure no work can be done in extension programme, for this active workers in this field should be encouraged so that they participate and enthusiastically remain active.

Principle of Evaluation: It is necessary to evaluate the extension work after a certain period so that merits and demerits of extension work can come to light and necessary changes be brought about. Evaluation generates confidence in people.

Principle of Leadership: The participation or inclusion of local leaders in extension programmes is the only criteria for assessing the success or failure of any extension work. Local leader is the best medium for dissemination of new ideas in rural area. According to Ensminger. "No Extension worker can be successful till he receives the cooperation of local leader of the village.

Principle of Learning by Doing: According to this principle farmers are encouraged to learn by doing the work themselves and by participating themselves. When a person does a work he gains practical knowledge and experiences the difficulties.

Principle of Need and Interest: Extension work is self-educative process. To make extension work more effective, it is necessary that it should be done according to peoples, needs and interest. Extension worker should take notice of local peoples' needs. As this will interest of the people which in turn will lead to their cooperation in extension work.

Principle of Neutrality: Extension worker should never take interest in local politics. If he will not behave in this manner then lots of difficulties in extension work will arise. Therefore, remaining neutral is more beneficial. Extension worker should never express his special affection or hatred towards any person.

Principle of Satisfaction: If people are not interested in extension work then there is no possibility that extension work can be carried on for a long time. In a democratic structure/set-up people cannot be run in a mechanized way. They should derive full satisfaction from extension work.

Principle of Trained Specialists: It is very difficult that extension personnel should be knowledgeable about all problems. Therefore, it is necessary that specialists should impart training to the farmers from time to time.

Principle of whole Family: In Extension Education, principle of whole family is of utmost importance. Family is a part of the community, so there should be development of each member of the family and extension education should be for each and every member of the family, i.e., boys, girls, male and female, then only success can be achieved.

Project of Local Area Development of Parliament Members: It has been often seen that the area from which the parliament members are elected, the people of the constituency come urging to them for their work. These work need financial assistance.

Rastriya Mahila Kosh: The establishment of Rastriya Mahila Kosh had been initiated with Rs. 31 crore as a fund. The main object of this programme was to give economic help in the form of credit to the women of poor class. The registration as a society has been done under the Society Act 1860. A board manages this kosh. Minister for Women and Children is the chairman of this board.

Reconstruction of Village: Work should be done for the development of education, health, transportation, electricity, water etc in villages, so as to encourage educated people to stay in villages. Arrangement should be made for providing security to the people in villages.

Rural Landless Employment Guarantee Programme (RLEGP): This project was running purposefully by the total help of the central government from 1983 in the rural area by providing the employment of 100 days work in a year to at least one member of landless labourer and to construct permanent properties, which may be helpful for the rural development.

Sampoorana Gramin Rozgar Yojna: The Sampoorana Grameen Rozgar Yojana (SGRY) was launched on 25 September 2001 by merging the on-going schemes of EAS and the JGSY with the objective of providing additional wage employment in the rural areas as also food security, beside the creation of durable .community assets in the rural' areas. The programme is self-targeting in nature with special emphasis on women, scheduled castes, scheduled tribes and parents of children withdrawn from hazardous occupations. While preference will be given to BPL families for providing wage employment under SGRY, poor families above the poverty line can also be offered employment under the programme.

Self-Help Groups (SHGs): Self-help groups provide mutual support among peers, especially women, in rural communities. Mutual support is a process by which people voluntarily come together to address common problems, especially in improving the livelihoods of group members, such as increasing access to education and health services. Self-help groups can also transition into specific economic activities of mutual interest that are culturally acceptable within rural communities.

Seva Gram: Mahatma Gandhiji was a great social worker. He knew very well that as long as Indian people are suppressed ,their society and their nation cannot progress. For ending this suppression, he began this welfare project "SEVA GRAM", establishing his Ashram in Vardha, in 1920. The main objective of this programme was to prevent the economic and social suppression of the people and to create the feeling of patriotism among them and they must think that this is their own country.

Shriniketan Project: Shriniketan is situated about 100 km. away from Calcutta (Kolkata) in West-Bengal State. This area was backward socially, economically and politically. Shri Ravindra Nath Tagore began this project of village development in this area with the help of sociologist Shri L.M. Hurst. Shri Tagore thought that if some villages were developed, the other villages will get inspiration and the programme of village development will spread all over the country and thus the whole country would be developed.

Small Farmers Development Agency (SFDA): In 1967, for the welfare of small farmers, this project was formed by the Government of India and was named "Small Farmers Development Agency". The registered Council implemented it.

Strategic Research and Extension Plan (SREP): Formulating a strategic research and extension plan involves identifying the farming systems and the resource base of farmers within a target area, as well as identifying the successes and failures of innovative farmers. It also involves the identification of problems and needs of farmers by using participatory rural appraisal (PRA) techniques and then analyzing all of this information using a SWOT (strengths, weaknesses, opportunities, and threats) analysis.

Sustainability: The underlying definition is the one adopted by the FAO in 1988, sustainable development is the management and conservation of the natural resource base, and the orientation of technological and institutional change in such a manner as to ensure the attainment and continued satisfaction of human needs for present and future generations. Such sustainable development (in the agriculture, forestry, animal and fisheries sectors) conserve land, water, plant and animal genetic resources, is environmentally non-degrading, technically appropriate, economically viable and socially acceptable.

Swarna Jayanti Gram Swarojgar Yojana (SGSY): The Government of India in the entire country introduced Swarna jayanti Gram Swarojgar Yojana (SGSY) from

1st April 1999. After the introduction of this Yojana the IRDP, TRYSEM, DWCRA, GKY, SITRA all emerged from this one project only. The main object of SGSY is to raise the poor people (self job workers) from below poverty line within three years by providing income-generating activities with the cooperation of government aid and bank loan.

Technology Assessment: Technology assessment for sustainable agriculture and rural development is defined here as a comprehensive approach to examine the actual or potential impact of technology applications on certain sustainability issues and second order consequences and to facilitate the development and use of technological interventions according to location-specific constraints and objectives.

Technology Transfer: Technology transfer is the process of disseminating new technologies and other practical applications that largely result from research and development (R&D) efforts in different fields of agriculture.

The basic Concept of Extension is that it is Education: Means production of desirable change in human behaviour, which includes knowledge, attitude and skill.

To Develop the Close Relations between Research Center and Agricultural Farm: A close coordination is to be developed between research center and agricultural farm. So that scientific information can be given to the farmers and their problems can be addressed to the scientific institution for solution.

Training and Visit (T&V) Extension: Training and Visit extension is based on classical management principles, including that (1) extension agents should have primary responsibility for carrying out extension functions; (2) extension should be closely linked with research; (3) training should be carried out on a regular and continuous timetable; (4) work should be time-bound; and (5) a field and farmer orientation should be maintained. This technology-driven approach was initially successful during the late 1970s and 1980s in disseminating the production management practices associated with Green Revolution wheat and rice varieties.

Training of Rural Youth for Self-Employment (TRYSEM): "Training of Rural Youth for Self Employment"- this programme has been structured in the form of co-operative project under the Integrated Village Development Project from 1979-80. Under this project, the marginal and small farmers living in the rural areas, the agricultural labourers and the rural young boys and girls whose age is between 18 to 35, belonging to the families of rural artisans and living below poverty line, are being trained and later settled in the self-businesses of local resources.

Transfer of Technology Programme (TOT): Today if we observe the economic progress of the country, it is evident that if the farmer is prosperous, the economic progress of the country will gain speed. For the prosperity of farmers, it is necessary that farmers obtain maximum production from the fields. This high yield can be possible only when the advantages (new practices) of research is practiced in fields of farmers in which, there may be improved seeds for more production of different crops or the latest developed crop-production technology.

Tribal and Hill Area Development Programme: This programme was introduced in the last years of "Fourth-five Year Plan" by the central government in 1973-74.

Value Chain: A value chain is an alliance of enterprises that collaborate "vertically" to achieve a more profitable position within a market. Vertically aligned means that both producers and essential companies are connected from one end of the primary production process (e.g., farmer's fields) through processing and then into the final marketing stages where consumers purchase a finished product.

Women's Development Corporation (WDC): The project was initiated in 1986-87 in different states of the country. The Women Development Corporation (WDC) performs the role of a motivator in the development of women. The main work is to provide the facility of credit card, to suggest about technology, to bring the articles made by women in the market, to support the Women Co-operative Institutions and to train them. This Women Development Corporation is working in Andhra Pradesh, Goa, Gujarat, Haryana, Himachal Pradesh, Jammu & Kashmir, Karnataka, Kerala, Maharashtra, Meghalaya, Odisha, Punjab, Tamil Nadu, U.P., West Bengal and other central territorial areas.

Women's Development Projects: On the basis of the census of 2001, the population of women and children is 49.87 crore (48.2%) and 86 lakh (15.42%) respectively, out of the total population of the country. After independence, many projects for women development are being launched in every five-year plan but in the beginning of 1980, in the sixth five year plan a new and separate Women and Child Development Project was initiated. For this, another department has been established under Human Resources Development Ministry. Its main work is to solve the problems of women and to prepare the action plan and to ensure the participation of women in the national development.

www.ingramcontent.com/pod-product-compliance
Ingram Content Group UK Ltd.
Pitfield, Milton Keynes, MK11 3LW, UK
UKHW021948270726
14060UKWH00002B/417